BAOFENG RADIO

YOUR SECRET WEAPON FOR SAFETY AND CONNECTIVITY

BRADLEY PRESTON

CONTENTS

ABOUT THE AUTHOR

BRADLEY PRESTON is an expert in radio and crisis communication, blending a background in law enforcement with practical crisis consulting experience.

As a respected speaker and consultant in law enforcement and communication strategies, Bradley is now sharing his experience and practical insights in navigating high-stakes situations to a wider audience.

INTRODUCTION

Imagine yourself on a trekking adventure, surrounded by the calm embrace of nature, when suddenly things go wrong, and you find yourself lost in the vast woods. Your phone signal fades, and the dense forest proves to be too much for your walkie-talkies. The traditional channels of contact are breaking down, leaving you isolated and alone.

Enter the Baofeng radio, your potential savior. In such circumstances, the Baofeng radio isn't just a tool; it will become your lifeline, a connection to the outside world that can turn a moment of adversity into a triumphant return to safety. In my line of work, I have dodged similar tricky situations more than once by carrying this simple gear. This is why I recommend *everyone* do the same.

In an era dominated by instantaneous digital connectivity, the importance of communication during emergencies is not highlighted much. We take communication for granted, assuming that our phones and the internet will always be there when needed. But what if they are not? What if unforeseen circumstances disrupt these lifelines, leaving us isolated and uninformed? It's a possibility we tend to overlook, yet it's a reality that unfolds more frequently than we might imagine.

But there is a straightforward, affordable option that can protect you from such scenarios when modern methods fail. For less than $30, a Baofeng Amateur Radio, also known as a Ham Radio, presents itself as an invaluable tool for establishing backup communications. It's a device that goes beyond the conventional notion of a two-way radio, becoming a lifeline that connects you with your local community, provides crucial weather alerts, delivers real-time updates from first responders, and serves as a beacon for help in times of need.

In a world where complex communication systems like cellphones and the internet rely on vast infrastructures, Ham Radio stands as a testament to individual responsibility. A simple setup—electricity, a radio, an antenna, and perhaps a few accessories—is all that's required for effective communication.

The rise in popularity of Baofeng Ham Radios can be attributed to their versatility, portability, and affordability. They have evolved into indispensable tools for amateur radio operators, outdoor enthusiasts, preparedness advocates (preppers), and individuals seeking reliable communication across diverse settings. At the forefront of this evolution is the UV–5RA, introduced in 2012, widely celebrated as the best Ham Radio for the money.

With a frequency range of 65 to 108 megahertz (MHz), high and low power levels, and programmable features, the UV–5RA offers a spectrum of capabilities. Its 128 programmable memory channels, rechargeable 1800 mAh Lithium-ion battery, and power-saving features make it a standout choice for all users, from novices to seasoned operators.

This book is your comprehensive guide to the Baofeng UV–5RA, an affordable entry-level radio that should find its place in every prepping family's toolkit. It will help you unlock the potential of the UV–5R, providing insights and knowledge to enhance your communication skills. Whether you're venturing into amateur radio, preparing for emergencies, exploring the outdoors, or simply expanding your communication horizons, the UV–5R stands ready to meet your needs.

This book aims to cater to both UV–5R users and those employing other Baofeng devices. Generalized instructions and tips provide a foundation, followed by procedures specific to the UV–5R model. The goal is to empower you to swiftly grasp and utilize your UV–5R, (or any Baofeng radio), offering a sound starting point for seamless integration into your communication toolkit.

As we dive deep into the exploration of Baofeng radios, we start with this first chapter, describing the origins of this fascinating device, and its multifaceted capabilities.

UNDERSTANDING THE BAOFENG PHENOMENON

"Communication is not just about transmitting information; it's about bridging gaps, creating connections, and sometimes, it's about the tools we choose to forge those links."

~ *Bob Edwards*

I have been a radio aficionado for a long time, always looking for the latest and most reliable devices to experiment with. I have worked with many different kinds of communication equipment, but very few of them impressed me as much as the Baofeng radio. I know there are so many devices nowadays that can do so many things, but Baofeng radios always feel unique because they fuse the simplicity of traditional communication with the advancements of the modern era. It's a small, cheap, versatile device that has *so many* features and functions. It's like having a mini radio station in your pocket!

In this chapter, we talk about the origins of this amazing radio device and its capabilities, and why I believe it to be one of the *best* radio tools *ever*.

INTRODUCTION TO THE BAOFENG BRAND

In the world of wireless communication, one brand has consistently pushed boundaries and redefined the standards: Baofeng. Born in 2001, with a vision to deliver value to the user and a commitment to innovation, Baofeng has traversed an impressive journey over the past 22 years.

In the early 2000s, a small factory of just over 30 people in Fujian, China, embarked on a mission that would revolutionize the world of handheld wireless communication. This was the birth of Baofeng Electronics Co. Ltd., a company that started with a simple vision: to create user value and innovate. The team at Baofeng was small but dedicated. Their decade's hard work paid off in May of 2010, when they launched their first dual-band LCD walkie-talkie. This innovative product rapidly increased the company's market share.

Baofeng's classic products, the UV–3R and UV–5R, received CE certification from the European Telecommunications Standards Committee in March of 2011, followed by RoHS (Restriction of Hazardous Substances in Electrical and Electronic Equipment) certification in April of the same year. This laid a solid foundation for Baofeng's presence in the European market.

However, perhaps the most important year for Baofeng was 2012. In 2012, Baofeng introduced the UV–5R, a handheld radio that would become a game-changer. This model was the first dual-band (VHF/UHF) radio from a Chinese firm to be commercially successful worldwide. Due to its low cost and ease of usage, radio amateurs and professionals worldwide have come to rely on it. The UV–5R was not without its challenges, though. It was only permitted for use on amateur radio and was not certified for use in the GMRS (General Mobile Radio Service) or FRS (Family Radio Service) in the United States due to its lack of FCC Part 95 certification. (Meacher 2019). Despite this, the UV–5R's popularity continued to grow, and Baofeng launched several other models based on the UV–5R technology.

What began as a modest factory with just over 30 people has blossomed into a high-tech enterprise, a powerhouse encompassing research and development, production, sales, and after-sales service of handheld wireless walkie-talkies and their accompanying accessories. They have more than 1,000 employees and a production base covering an area of more than 30,000 square meters, (Baofeng, n.d.).

Since its founding, Baofeng has narrowed its emphasis to radio communication research and development, remaining committed to leading the way in product innovation.

Baofeng's journey was not always smooth sailing. The UV–5R attracted the attention of multiple telecommunications regulators due to issues primarily relating to frequency interference. It was banned from sale and use in several countries, including Switzerland, Germany, Poland, and South Africa (Baofeng, n.d.) However, Baofeng did not let these challenges deter them. They continued to innovate and improve their products, ensuring they met the necessary standards and certifications. Their perseverance paid off, and today, Baofeng radios are sold all over the world.

Baofeng's story is a testament to the power of innovation, perseverance, and a commitment to creating user value. From that small factory in Fujian to a global leader in wireless communication, Baofeng's journey is truly inspiring.

And so, as you switch on your Baofeng radio, remember the journey it has taken. From the minds of a dedicated team in a small factory in China, to your hands, it's not just a device; it's a story of technical creativity and determination.

WHY ARE BAOFENG RADIOS SO POPULAR?

At the heart of Baofeng's widespread acceptance is its unmatched affordability. In a world where communication devices often come with quite a hefty price tag, Baofeng radios break the mold. They offer a cost-effective solution, bringing two-way radio communication within reach of a diverse user base. This affordability factor has democratized access to radio communication, enabling more individuals to integrate these devices into both personal and professional spheres.

But these radios go beyond just being cheap. There are a plethora of reasons why everyone requiring a Ham Radio usually chooses Baofeng over any other brand.

Here are some features that set it apart from other radio brands:

- **Dual band and dual display:** The Baofeng radio can operate on both the VHF and UHF bands, covering a wide range of frequencies from 136 to 174 MHz and from 400 to 520 MHz. It can also display two frequencies or channels simultaneously on its LCD screen, allowing the user to switch between them easily.

- **Programmable and customizable:** The Baofeng radio can be programmed manually using the keypad, or through a computer using the CHIRP application and a USB cable. The user can customize the radio settings, such as the frequency, channel, power, squelch, scan, and more. The user can also assign different functions to the keys and buttons, such as the PTT, A/B, FM, and MONI.

- **Compatible and versatile:** The Baofeng radio is compatible with various accessories, such as antennas, batteries, chargers, earpieces, microphones, and speakers. The user can enhance the performance and functionality of the radio by using different accessories according to their needs and preferences. The radio is also versatile and can be used for various purposes, such as: Ham Radio, emergency communication, outdoor activities, and more. This versatility caters to a broad spectrum of user needs, making Baofeng radios indispensable for various scenarios.

- **Durable and portable:** The Baofeng radio is made of high-quality materials and has a solid and compact design. It is resistant to dust, water, and shock, and can withstand harsh environments and conditions. The radio is also lightweight and easy to carry, and comes with a belt clip and a hand strap for convenience.

By making radio communication accessible to a broader audience, Baofeng has not only transformed the way we communicate but has also prompted crucial discussions about regulatory compliance, and the responsible use of these powerful devices.

The Baofeng phenomenon is a testament to the brand's ability to innovate and redefine the norms of radio communication in the modern era.

HANDHELD TRANSCEIVERS AND THEIR ROLE IN MODERN COMMUNICATIONS

With the wide adoption of cellphones and the internet, communication has become easier than ever. Handheld transceivers, also known as walkie-talkies, two-way radios, or HTs (handie-talkies) are largely replaced by phones, cellphones, or internet-based services. Yet, they possess some advantages we use every day.

Handheld transceivers are much cheaper than any modern tech in the market. These devices can be bought for as low as $25 and are designed to withstand harsh conditions, such as dust, water, shock, and temperature extremes. Some models are even military-grade and meet the standards of IP67 or MIL-STD-8103. This makes them ideal for use outdoor and in rugged environments, where other devices may fail or break easily. These transceivers can beat any smartphone when it comes to durability and rigidity, because the phones we use are not meant for such extreme conditions.

These devices are also *independent* of any infrastructure, such as power lines, cell towers, or satellites, which may be unavailable or unreliable in remote areas or during disasters. When you are hiking or camping with your friends or family, how often do you find yourself with *no* network or cellular coverage? If you are a travel enthusiast like me, I would assume it happens pretty often. Handheld transceivers do not rely on *anything*. This means that they can work even in areas where there is no electricity, cellular coverage, or internet access. This also means that they can function during disasters, such as earthquakes, floods, fires, or terrorist attacks, when the infrastructure may be damaged or overloaded. This gives them an edge over other devices that rely on the availability and reliability of the infrastructure.

With the widespread use of social media, the internet and smartphones, our privacy has become a major concern. We are at constant risk of exposing our personal information to Big Tech even when we don't consent to it. And honestly, these technologies have entangled us in such a way that there is no way around, unless you choose to use something that cannot be easily intercepted or monitored by others. Yes, these handheld transceivers use dedicated frequencies that are assigned by the authorities or agreed upon by the users. These frequencies are not shared with other users or services, and are not easily accessible by others. Some models also use encryption or scrambling techniques to protect the content of the communication from anyone eavesdropping or jamming. This makes them more secure and private than other devices that use public or common frequencies that can be intercepted or monitored by others.

Handheld transceivers have a long history of development and use, dating back to the Second World War, when they were first invented by various engineers and scientists. Now, in addition to professional use, they are widely being used for personal and recreational purposes, such as communication between family members, friends, co-workers, emergency responders, hobbyists, and enthusiasts. While handheld radios might seem like an outdated device with the widespread use of smartphones, it is a no-brainer when it comes to communication during emergencies and disasters. In the aftermath of natural or man-made disasters, such as earthquakes, floods, fires or terrorist attacks, handheld devices outplay *any* other modern devices.

One of the most tragic examples of such a disaster was the massive earthquake that struck Haiti on January 12, 2010, killing more than 200,000 people and displacing millions more. The quake also destroyed much of the infrastructure, including power lines, cell towers, and internet services, making communication extremely difficult and hampering rescue efforts. In this dire situation, handheld transceivers proved to be invaluable devices for the survivors and the rescuers alike, (Sturcke 2010).

Many people who were trapped under the rubble of collapsed buildings used their walkie-talkies to call for help and to stay in contact with their loved ones. Some of them were able to survive for days or even weeks by drinking water from pipes or toilets and eating whatever they could find, while keeping their hopes alive by listening to the voices of their rescuers or family members. For example, Darlene Etienne, a 16-year-old girl, was pulled out alive after 15 days under the debris of a house near her college, thanks to her handheld transceiver that allowed her to call for help and to be located by a French rescue team, (Sturcke 2010).

The Haiti earthquake of 2010 was one of the worst humanitarian disasters in history, and it showed the importance, and the limitations, of communication in such a crisis. Handheld transceivers were among the few devices that could function and provide a lifeline for the survivors and the rescuers.

On a personal note, handheld transceivers can be used by individuals for recreation and entertainment by hobbyists such as hikers, campers, hunters, and amateur radio operators. I go camping every now and then with my children and we always take our handheld transceivers with us. It not only ensures security in case of any mishaps, but also acts as a great recreational tool. My kids love a game of hide and seek and our handheld transceivers add more excitement to the whole game!

We divide into two teams, and each team has a transceiver of their own. One team hides somewhere in the forest, and the other team tries to find them using clues and directions from the radio. In the midst of the forest, this is the only piece of tech that ensures secure connection and makes sure *none* of us get lost. It also makes the whole experience much more wholesome. We have also used the UV-5RA to communicate with each other and coordinate our activities in a theme park. We could tell each other where we were, what rides we wanted to go on, and where to meet for lunch. We also used them to prank each other and make jokes. The Baofeng radio is a simple device that emulates the experience of being in touch, without succumbing to notifications and message alerts.

While handheld transceivers may seem to be overshadowed by the prevalence of smartphones and internet-based communication, they hold significant advantages in different circumstances. The landscape of communication technology continues to evolve but handheld transceivers persist as resilient and versatile devices, proving their worth in both critical situations and everyday recreational pursuits.

Now, as you know the history and significance of a Baofeng radio, it's time to learn the basics of your first Baofeng radio.

2

THE BASICS: YOUR FIRST BAOFENG RADIO

"The radio is the most important tool in the world. It is the bridge between people, between cultures, between nations."

~ Nelson Mandela

I once worked with Maria, an enthusiastic hiker who told me an incredible story about one of her first experiences with a Baofeng radio.

Initially, the multitude of buttons on the device seemed daunting, and she found herself clumsily navigating through the manuals, expecting nothing more than static.

During a solo expedition, an unexpected crackle from the speaker startled her – it was a distress call from a fellow hiker. With her heart pounding and hands trembling, Maria mustered the courage to navigate the channels of her Baofeng.

In that critical moment, she transformed from a solitary hiker into a lifeline, guiding the lost individual to safety. That day, amidst the echoes of the mountain, Maria's fear turned into awe. Her mountaintop experience changed forever, all thanks to her first Baofeng radio.

When people start using a Baofeng radio for the first time, they are faced with multiple challenges. *How do I turn it on? How do I change the frequency? How do I program the channels?* Users have a lot of questions popping up in their mind on first using the device.

It's okay to feel confused and overwhelmed due to the device's multifaceted character. But I don't want you to feel at all discouraged. So, in this chapter, I will guide you through the basics of choosing the right model for your needs, understanding the physical layout of the radio, and setting it up for the first time. By the end of this chapter, you will be able to use your Baofeng radio with confidence and ease.

DIFFERENT TYPES OF BAOFENG RADIO

Baofeng radios' appeal lies in their affordability, user-friendly design, and versatility across various frequency bands and modes. However, it's important to note that not all Baofeng radios are created equal. Let's dive into the distinct features and characteristics of some of the standout models.

1. Baofeng UV–5RA Dual Band Ham Two-Way Radio

The Baofeng UV–5RA caters to beginners and casual users with its compact, lightweight design. Operating on both

VHF and UHF frequencies, it supports analog transmissions and features a built-in FM radio, flashlight, and alarm function. While it may lack the power of the BF-F8HP (discussed later), the UV–5RA remains an affordable option for short-range communication and leisure activities. I personally recommend this piece of tech to anyone who is looking for a pocket walkie talkie for personal and family use.

The Baofeng UV-5RA has a notable legacy in the world of handheld radios. However, it's important to note that the original UV-5RA has been discontinued from production for quite some time now. This model, which has been the cornerstone of Baofeng radios, is no longer being manufactured, marking the end of an era. Despite its discontinuation, the UV-5RA's influence and impact remain significant in the field.

After the UV-5RA's discontinuation, Baofeng introduced the BF-F8HP, the third generation of this model. This new model can be seen as a new generation version of the UV-5RA, carrying forward its legacy while introducing new features and improvements. Despite the change in model name, the UV-5RA name continues to be widely recognized and used.

In this book, I will continue to refer to the UV-5RA, as its functionalities remain relevant even for subsequent models.

Pros:

- Compact and lightweight design
- Easy to use and program
- Budget-friendly price
- Comes with an earpiece

Cons:

- Low output power (4 watts) for limited range and clarity
- Standard antenna for average reception and transmission
- No digital mode or encryption support
- May require a license to operate legally

2. Baofeng BF-F8HP Dual Band Two-Way Radio

The Baofeng BF-F8HP stands out as one of the most powerful and advanced radios in the Baofeng lineup. It's essentially an upgraded version of the popular UV-5RA model, boasting twice the output power at 8 watts, a larger 2000 mAh battery, an improved V-85 antenna, and an enhanced user manual. Operating on both VHF and UHF bands, it supports a variety of analog transmissions and features a built-in FM radio, flashlight, and alarm function. The user-friendly interface, backlit LCD display, and programmable options make it an ideal choice for those requiring high-performance and reliable communication, especially in long-range and emergency situations.

Pros:

- High output power (8 watts) for extended range and clarity
- Large battery (2000 mAh) for longer battery life
- Improved antenna (V-85) for enhanced reception and transmission
- User-friendly interface and programmable options
- Compatibility with UV-5RA accessories

Cons:

- Not waterproof or dustproof
- No digital mode or encryption support
- May require a license to operate legally

3. Btech UV-5X3 Tri-Band Radio

The Btech UV-5X3 stands out as a unique and versatile option, operating on three frequency bands: VHF, UHF, and the exclusive 1.25M band for US amateur radio operators. This tri-band capability provides more frequency options with less congestion. Additionally, it supports FM radio and NOAA weather alerts, making it a feature-rich choice. While it has a lower output power than the BF-F8HP, the UV-5X3 offers a compelling array of functions for users looking to explore the 1.25M band.

Pros:

- Tri-band radio for more frequency options and less congestion
- Supports FM radio and NOAA weather alerts
- Comes with two antennas for different bands
- User-friendly interface and programmable buttons
- Comes with an earpiece

Cons:

- Low output power (5 watts) for limited range and clarity
- Smaller battery (1500 mAh) for shorter battery life
- Requires a license to operate legally

- More expensive than other Baofeng radios

The choice of the right Baofeng radio ultimately depends on the unique needs and preferences of the user. Each model, whether it's the compact and budget-friendly UV5RA, the high-performance BF-F8HP, or the versatile tri-band UV-5X3, caters to a specific set of requirements. Whether you prioritize simplicity, power, or versatility, Baofeng radios provide a diverse range of options. It's a matter of aligning the specific features and capabilities of each model with individual preferences, ensuring that the chosen radio meets the unique communication needs of the user.

NAVIGATING THE PHYSICAL LAYOUT OF A BAOFENG RADIO

Now that you have an understanding of different types of Baofeng radios, it's time we talk about the physical layout of the radio. Unlike smartphones, the layout of these devices can be a little overwhelming when you are just starting out. There are many buttons, each with its own function. If you do not familiarize yourself with these buttons and keys, you will not be able to utilize this tool to its fullest potential. Let me specifically describe the Baofeng UV-5RA, because I believe it to be the best radio for beginners.

• **Antenna:** The antenna is the part that transmits and receives radio signals. It is located at the top of the radio and can be screwed on or off. The typical SMA-Female antenna that comes with the UV-5RA Baofeng radio is about 15 cm long and 1 cm in diameter. Depending on your requirements and tastes, you may also use other suitable antennas, such as longer or shorter ones. Simply align the radio's and antenna's threads and twist them clockwise until they are securely fastened to complete the installation process. To remove the antenna, twist it counterclockwise and pull it out, (Baofeng, n.d.).

• **Battery:** The battery is the part that powers the radio. It slides on and off and is situated at the rear of the radio. A 1800 mAh rechargeable Lithium-ion battery with a 7.4 V voltage is included with the UV-5RA Baofeng radio. Other suitable batteries, such as ones with a larger or lower capacity, may also be used, based on your requirements and preferences. To install the battery just line up the radio's and the battery's grooves and slide them together until they click. To remove the battery, press the release button at the bottom of the radio and slide the battery out, (Baofeng, n.d.).

• **Buttons:** The buttons are the parts that control the functions and settings of the radio. They are located on the front, side, and top of the radio and have different shapes and labels. The UV-5RA Baofeng radio has the following buttons:

○ *Power/Volume Knob:* The power/volume knob is the part that turns the radio on or off and adjusts the volume level. It is a circular object with a red dot at the top of the radio. Switch the knob clockwise until you hear a prompt spoken and a click to switch on the radio. Switch the knob counterclockwise until you hear a click, and the screen turns black, to switch off the radio. When the radio is on, turn the knob clockwise or counterclockwise to change the volume level.

○ *PTT Button:* The PTT (Push-To-Talk) button is the part that activates the transmission mode. It is located on the left side of the radio and has a rectangular shape and a red label. To transmit, press and hold the PTT button while speaking into the microphone. To stop transmitting, release the PTT button.

○ *MONI Button:* The MONI (Monitor) button is the part that activates the monitor mode. It is located underneath the PTT button on the left side of the radio, and it is rectangular in form with a black label. When the radio is on, press and hold the MONI button to monitor. By doing this, you'll be able to hear all signals on the current channel or frequency and the squelch will open. To stop monitoring, release the MONI button.

○ *CALL Button:* The CALL (Call) button is the part that activates the call mode. It is rectangular in design, with a black label, and it is located on the left side of the radio, above the PTT button. When the radio is on, press and hold the CALL button to make a call. This will notify other radios on the same frequency or channel by sending out a preprogrammed tone or code. Press and release the CALL button once more to end the call.

○ *A/B Button:* The A/B (A/B) button is the part that switches between the two display lines: A and B. It is spherical in form with a black label, and it sits on the front of the radio underneath the display. When the radio is on, push and release the A/B button to switch. By doing this, the active display line will alternate between A and B. An arrow on the left side of the display indicates the current display line.

○ *BAND Button:* The BAND (Band) button is the part that switches between the two frequency bands: VHF and UHF. It is spherical in form with a black label, and it sits on the front of the radio underneath the display. While the radio is on, push and release the BAND button to switch. By doing this, the active display line's frequency band will be switched between VHF and UHF. On the right side of the display, a letter denoting the frequency band is displayed: V for VHF and U for UHF.

○ *VFO/MR Button:* The VFO/MR (Frequency Mode/Memory Mode) button is the part that switches between the two operation modes: frequency mode and memory mode. It is spherical in form with a black label, and it is placed on the front of the radio underneath the display. While the radio is on, push and release the VFO/MR button to switch. By doing this, you may switch the active display line's operating mode between memory and frequency modes. A sign on the left side of the display denotes the operating mode: a channel number for memory mode, and a frequency symbol for frequency mode.

○ *MENU Button:* The MENU (Menu) button is the part that accesses the menu settings. It is spherical in form with a black label, and it sits on the front of the radio underneath the display. When the radio is on, press and hold the MENU button to access. This will bring up the menu and show the first option on the menu, SQL (Squelch Level). You may either push and release the EXIT or MENU buttons to leave.

○ *EXIT Button:* The EXIT (Exit) button is the part that exits the menu settings or the keypad lock. It is spherical in shape, has a black label, and is located underneath the display on the radio's front. To quit the radio when it is in menu or keypad lock mode, press and hold the quit button. This will put an end to the menu mode or keypad lock mode and return to standard mode.

○ *Keypad:* The keypad is the part that inputs numbers, letters, symbols, and commands. It is located on the front of the radio, below the display, and has 16 keys with different labels. The keypad has the following functions:

▪ Numeric Keys: These are the keys that allow you to enter numbers between 0 and 9. They feature black labels and square forms, and they are on the keypad. When the radio is in either the frequency or memory mode, enter data by pressing and releasing the numeric key. As a result, the current display line's number will be entered.

▪ Star Key: The decimal point or negative sign is entered with the star key. It is square-shaped with a black label, and you may find it on the keypad under the 7 key. When the radio is in either the frequency or memory mode, enter data by pressing and releasing the star key. By

doing this, the active display line will be
updated with the decimal point or the negative
sign.

- Pound Key: The pound key is the key that
  activates the keypad lock or the scan function.
  It is located on the keypad, below the 9 key, and
  has a square shape and a black label. To
  activate, press and hold the pound key for
  about 2 seconds while the radio is on. This will
  toggle the keypad lock or the scan function on
  or off. The keypad lock and the scan function
  are indicated by a key symbol and a scan
  symbol respectively on the right side of the
  display.

- UP Key: The UP (Up) key is the key that
  increases the value or selects the next item. It is
  located on the keypad, above the 2 key, and has
  a triangular shape and an up arrow label. To
  increase or select, press and release the UP key
  while the radio is in the frequency mode, the
  memory mode, or the menu mode. This will
  increase the value or select the next item on the
  active display line.

- DOWN Key: The DOWN (Down) key is the key
  that decreases the value or selects the previous
  item. It is located on the keypad, below the 8
  key, and has a triangular shape and a down
  arrow label. To decrease or select, press and
  release the DOWN key while the radio is in the
  frequency mode, the memory mode, or the
  menu mode. This will decrease the value or

select the previous item on the active display line.

- A Key: The A (A) key is the key that inputs the letter A or the plus sign. It is located on the keypad, above the star key, and has a square shape and a black label. To input, press and release the A key while the radio is in the frequency mode or the memory mode. This will input the letter A or the plus sign on the active display line.

- B Key: The B (B) key is the key that inputs the letter B or the equal sign. It is located on the keypad, above the o key, and has a square shape and a black label. To input, press and release the B key while the radio is in the frequency mode or the memory mode. This will input the letter B or the equal sign on the active display line.

- C Key: The C (C) key is the key that inputs the letter C or the slash sign. It is located on the keypad, above the pound key, and has a square shape and a black label. To input, press and release the C key while the radio is in the frequency mode or the memory mode. This will input the letter C or the slash sign on the active display line.

- D Key: The D (D) key is the key that inputs the letter D or the reverse slash sign. It is located on the keypad, below the 3 key, and has a square shape and a black label. To input, press and

release the D key while the radio is in the
frequency mode or the memory mode. This will
input the letter D or the reverse slash sign on
the active display line, (Baofeng, n.d.).

STEP-BY-STEP GUIDE FOR INITIAL SETUP

Now that you have a better understanding of the whole
layout, it's time we talked about how you can start *using*
your first Baofeng radio.

Installing the Battery Pack and Belt Clip: The
Baofeng UV-5RA comes equipped with a convenient
battery pack that is easily attached to the back of the radio
using a secure clip. Follow these simple steps for installing
and removing the battery pack:

To Remove the Battery Pack:

1. Locate the clip on the back of the radio.
2. Press the clip down.
3. Slide the battery pack off.

To Install the Battery Pack:

1. Align the metal contacts on the battery pack with
   those on the radio.
2. Slide the battery pack up until it clicks into place.

Belt Clip Attachment:

1. Use the provided belt clip for hands-free carrying.
2. To attach, align the holes on the clip with the
   screws on the battery pack.

3. Tighten the screws using a screwdriver.
4. To remove, loosen the screws and slide the clip off.

Power and Volume: Before you try to turn on the radio, *ensure the battery and antenna are attached.*

Turning On:

1. Rotate the volume/power knob clockwise until a "click" is heard.
2. The radio powers on with a double beep, and the display/backlight activates.

Quick Tip: Hold down specific keys during startup for additional information:

- [3 SAVE]: Firmware version.
- [6]: Manufacturing date and hardware revision number.

Turning Off:

1. Turn the volume/power knob counterclockwise until a "click" is heard.
2. The display and backlight will turn off.

Adjusting Volume:

1. Increase volume by turning the knob clockwise.
2. Decrease volume by turning the knob counterclockwise (be cautious not to turn it too far to avoid turning the radio off).

Quick Tip: Utilize the monitor function, enabled from

the Monitor key (below the PTT button), for easy volume adjustment by tuning it to the un-squelched static.

Making Calls: Transmitting with your Baofeng UV-5RA is pretty straightforward.

1. **Press and Hold PTT Button:**

a. Locate the button on the side of the radio body.
b. Press and hold the Push-To-Talk (PTT) button to transmit.

2. **Release PTT Key:**

a. Once your message is delivered, release the PTT key.
b. The transceiver will seamlessly shift back into receive mode.

Quick Tip: For optimal transmission, adhere to the following guidelines:

- Allow 1 to 2 seconds after pressing the PTT button before speaking.
- Maintain a distance of 1 to 2 inches from the microphone.
- Speak slowly and distinctly for clear communication.

Channel Selection:

Frequency (VFO) Mode: In Frequency (VFO) mode, you have the flexibility to fine-tune your communication by

adjusting the frequency within the designated band. Follow
these steps:

1. **Adjusting Frequency:**

a. Access Frequency (VFO) mode by pressing the
[VFO/MR] key.
b. Use the [UP ARROW] and [DOWN ARROW]
keys to increase or decrease the frequency.
c. Each press of the buttons corresponds to the
frequency step set on your transceiver.

2. **Direct Frequency Input:**

Input frequencies directly on the numeric keypad
with kilohertz accuracy.

*Note: The radio rounds up to the nearest frequency
based on your preset frequency step.*

Example (assuming a 12.5kHz frequency step):

a. Press [VFO/MR] to switch to Frequency (VFO)
mode.
b. Press [A/B] until [UP] appears next to the upper
display (display A).
c. Press [BAND] to select VHF (136-174 MHz).
d. Enter [1] [4] [7] on the numeric keypad.
e. Enter [6] [8] [8] on the numeric keypad.

Channel (MR) Mode:

Navigating Programmed Channels:

a. Switch to Channel (MR) mode when channels
are programmed into your Baofeng UV-5RA.
b. Use the [UP] and [DOWN] keys to navigate
between your programmed channels.

Quick Tip: If your programmed channels have the transmit
power set to "Low," you can temporarily boost it to "High"
power by pressing the [# KEY IMAGE] key.

Understanding both Frequency (VFO) and Channel (MR)
modes equips you with the tools to seamlessly communicate
with your Baofeng UV-5RA radio. Whether you prefer the
precision of adjusting frequencies in real-time or the
convenience of navigating through programmed channels,
these modes provide the flexibility you need for a smooth
and effective radio experience.

Navigating the Menu: Here's a step-by-step guide to
using the menu efficiently:

1. **Accessing the Menu:**

   a. Press the [MENU] key to enter the menu system.

2. **Navigation:**

   a. Use the [UP] and [DOWN] keys to scroll
   through menu items.

3. **Selecting Menu Items:**

   a. Once you find the desired menu item, press
   [MENU] again to select it.

4. **Adjusting Parameters:**

a. Use the [UP] and [DOWN] keys to select the desired parameter.

b. To confirm your selection, press [MENU]—this saves your setting and returns you to the main menu.

c. To cancel changes, press [EXIT]—this resets the menu item and exits the menu.

Exiting the Menu:

At any point, press the [EXIT] key to leave the menu.

Quick Tip: Every menu item has a numerical value associated with it, and the keypad is designed for easy access to the most common functions. Here is how you can use a shortcut to access it.

This chapter was all about understanding the key functions of a Baofeng radio. But this is just the beginning. There is a vast amount of things you can do with this simple device, and you've just taken your first steps into understanding the foundations. In the next chapter we are going to learn about programming a Baofeng radio, the understanding of which will highly improve your overall experience of using a pocket radio.

3

PROGRAMMING MADE EASY

"The secret of war lies in the communications."

~ *Napoleon Bonaparte*

Programming your Baofeng radio may seem like a daunting task, especially if you are new to the world of amateur radio. You may have heard of terms like frequency, channel, repeater, and privacy code, but you may not know how to use them, or what they mean. You may have also heard of software like CHIRP, but you may not know how to install it, connect it to your radio, or use it to program your radio.

In my consulting practice, I notice users can often be frustrated when first coming to grips with their Baofeng radio. That's why in this chapter, I will share with you my experience and tips on how manually program your Baofeng radio in case you don't have access to a computer or a cable. I will also show you how to use CHIRP to set up your radio. Whether you are a beginner or an expert, you will find something useful and interesting in this chapter. So, let's get started.

SIGNALING SYSTEMS

DTMF, CTCSS, and DCS are three types of tone signaling systems that are used in two-way radio communication. They are used for different purposes, such as activating repeaters, controlling repeaters, accessing IRLP links, making autopatch calls, and filtering out unwanted signals.

DTMF

DTMF stands for Dual Tone Multi-Frequency. It is also known as Touch-Tone, because it uses the same tones as the telephone keypad. DTMF uses two sine waves of different frequencies to produce 16 different tones. Each tone corresponds to a key on the keypad, such as 0-9, *, #, A-D. DTMF is an "in-band" signaling system, which means that the tones are audible and transmitted along with the voice.

DTMF is commonly used for remote control. A common example would be where the repeater is activated by sending out a DTMF sequence (usually a simple single-digit sequence). The Baofeng UV–5R has a full implementation of DTMF, including the A, B, C, and D codes. The numerical keys, as well as the [*SCAN] and [# KEY IMAGE] keys, correspond to the matching DTMF codes as you would expect. The A, B, C, and D codes are located in the [MENU], [UP], [DOWN] and [EXIT] keys respectively. QUICK TIP: Having the keypad lock enabled, you are still able to send DTMF tones without having to unlock the radio.

To send a DTMF tone, you need to be in the frequency mode (VFO) or the channel mode (MR) and press the PTT button. Then, you can press the keypad keys to send the corresponding tones. You can also press and hold the [# KEY IMAGE] key to send a continuous tone. To stop sending tones, you can release the PTT button or press the [EXIT] key.

You can also set up a DTMF memory, which allows you to store and recall a sequence of up to 16 DTMF tones. This can be useful for frequently used commands or codes. To set up a DTMF memory, you need to press the [MENU] key and then the 25 key. Then, you can enter the DTMF sequence using the keypad keys. You can also use the [UP] and [DOWN] keys to select a pre-stored DTMF memory. To save the DTMF memory, you need to press the [MENU] key again. To exit the DTMF memory mode, you need to press the [EXIT] key.

To use a DTMF memory, you need to press the [MENU] key and then the 26 key. Then, you can use the [UP] and [DOWN] keys to select the DTMF memory you want to use. To send the DTMF memory, you need to press the PTT button. To exit the DTMF memory mode, you need to press the [EXIT] key.

CTCSS

CTCSS stands for Continuous Tone-Coded Squelch System. It is also known as tone squelch or PL (Private Line). CTCSS is a "sub-audible" signaling system, which means that the tones are not easily heard by human ears and are usually filtered out before reaching the speaker or headphone. CTCSS works by adding a low-level tone to the voice, and muting the receiver unless it detects the same tone. CTCSS is mostly used for activating repeaters and filtering out unwanted signals on a shared channel.

CTCSS uses continuous tones below 300 Hz, and there are 50 standard tones. CTCSS tones range from 67 to 257 Hz.

DCS

DCS stands for Digital-Coded Squelch. It is also known as DPL (Digital Private Line) or CDCSS (Continuous Digital-Coded Squelch System). DCS is also a "sub-audible" signaling system, which means that the code words are not heard by human ears and are usually filtered out before reaching the speaker or headphone. DCS works by adding a low-level code word to the voice, and muting the receiver unless it detects the same code word. DCS is mostly used for activating repeaters and filtering out unwanted signals on a shared channel.

DCS uses 23-bit code words that are modulated at 134.4 Hz. There are 104 standard code words.

MANUAL PROGRAMMING

There are two ways to program a Baofeng radio UV–5R: using the keypad of the radio itself, or through computer software called CHIRP. However, CHIRP requires a programming cable that connects the radio to the computer, and it may not be available or convenient for some users. (I'll cover CHIRP programming in a separate section on the next page.)

Manual programming is useful for users who want to program a few channels on the go, or who want to learn how the radio works. Manual programming can be done in two modes: frequency mode and channel mode. Frequency mode allows users to enter and modify frequencies directly, while channel mode allows users to select and use the pre-programmed channels.

To program a Baofeng radio UV–5R manually, users need to know the following information:

- The frequency of the channel they want to program, such as 146.52 MHz.
- The frequency band of the channel, either VHF (136-174 MHz) or UHF (400-520 MHz).
- The CTCSS or DCS tone of the channel, if any. CTCSS and DCS are sub-audible tones that are used to access repeaters or filter out unwanted signals. CTCSS stands for Continuous Tone-Coded Squelch System, and DCS stands for Digital-Coded Squelch. CTCSS and DCS tones are expressed in hertz (Hz) or codes, respectively. For example, a CTCSS tone of 100 Hz or a DCS code of 023.
- The offset and direction of the channel, if it is a repeater channel. A repeater is a station that receives a signal on one frequency and retransmits it on another frequency, usually with higher power and wider coverage. A repeater channel uses two frequencies: one for receiving (input) and one for transmitting (output). The difference between the two frequencies is called the offset, and it can be positive or negative. For example, a repeater channel with an input frequency of 146.52 MHz and an output frequency of 146.94 MHz has a positive offset of 0.42 MHz.

Once users have the above information, they can follow these 10 steps to program a Baofeng radio UV–5R manually:

1. Turn on the radio and press the [VFO/MR] button to enter the frequency mode. The display should show a frequency number, such as 146.520, and no channel number on the right side.

2. Press the [A/B] button and choose the A side (upper display). The A side must be used to program channels into the radio. Programming data entered on the B side (lower display) will not be saved.

3. Press the [BAND] button to choose the frequency band. Toggle the [BAND] button to switch between VHF and UHF. The display should show the letter V or U on the left side, indicating the band. Make sure to choose the correct band for the frequency you want to program. If the wrong band is chosen, the radio will cancel the operation.

4. Disable the TDR (Dual Watch/Dual Standby) function. The TDR function allows the radio to monitor two frequencies or channels simultaneously, but it may interfere with the programming process. To disable the TDR function, press the [MENU] button, then press the number 7 on the keypad, then press the [MENU] button again. Use the up and down arrow keys to select OFF, then press the [MENU] button to confirm and save. Press the [EXIT] button to return to the frequency mode.

5. Enter the frequency of the channel you want to program. Use the keypad to enter the frequency number, such as 146520. The display should show

the frequency you entered. If you make a mistake, press the [*SCAN] button to delete the last digit, or press the [EXIT] button to cancel the operation and start over.

6. Optional—Enter the CTCSS or DCS tone of the channel, if any. To enter a CTCSS tone, press the [MENU] button, then press the number 13 on the keypad, then press the [MENU] button again. Use the keypad or the up and down arrow keys to enter or select the CTCSS tone in hertz, such as 1000. Press the [MENU] button to confirm and save. Press the [EXIT] button to return to the frequency mode. To enter a DCS code, press the [MENU] button, then press the number 12 on the keypad, then press the [MENU] button again. Use the keypad or the up and down arrow keys to select the DCS code, such as 023. Press the [MENU] button to confirm and save. Press the [EXIT] button to return to the frequency mode.

7. If the channel is a simplex channel, skip to step 9. If the channel is a repeater channel, continue to step 8.

8. Enter the offset and direction of the repeater channel. To enter the offset, press the [MENU] button, then press the number 25 on the keypad, then press the [MENU] button again. Use the keypad or the up and down arrow keys to enter or select the offset in megahertz, such as 0.600. Press the [MENU] button to confirm and save. Press the [EXIT] button to return to the frequency mode. To enter the direction, press the [MENU] button, then press the number 26 on the keypad, then press the [MENU] button again. Use the up and down arrow keys to select the direction, either + or

-. Press the [MENU] button to confirm and save. Press the [EXIT] button to return to the frequency mode.

9. Assign the frequency to a channel. Press the [MENU] button, then press the number 27 on the keypad, then press the [MENU] button again. Use the keypad or the up and down arrow keys to enter or select the channel number you want to assign the frequency to, such as 001. Press the [MENU] button to confirm and save. Press the [EXIT] button to return to the frequency mode. The display should show the channel number on the right side, indicating that the channel has been programmed successfully.

10. Repeat steps 1 to 9 to program more channels, or press the [VFO/MR] button to switch to the channel mode and use the programmed channels

PROGRAMMING WITH CHIRP

I highly suggest the utilization of CHIRP software for your amateur radio programming needs. As a complimentary, open-source platform, CHIRP offers extensive support for a wide array of manufacturers and models. It also provides a versatile interface that can interact with diverse data sources and formats, enhancing its usability and adaptability.

CHIRP, which stands for Compressed High Intensity Radar Pulse, is a signal in which the frequency increases (up-chirp) or decreases (down-chirp) with time. It is commonly applied to sonar, radar, and laser systems, as well as other applications such as spread-spectrum communications.

Programming your amateur radio involves a few steps:

Downloading the Appropriate Files: Before you can start setting up your radio station, you need to download the necessary files. This includes the software for your radio station, which could be something like CHIRP, and any other files necessary for the operation of your station. You can download CHIRP from its official website. The CHIRP team recommends using the latest build available. Since the software is offered as open-source, it's free of charge. It's important to note that you do not need to uninstall an existing version of CHIRP before installing a newer one, (Smith, n.d.).

Setting Up Your Server: Once you have downloaded CHIRP, you can start setting it up. Here are the steps to follow:

1. Start CHIRP.
2. Click the Radio menu and choose Download from Radio.
3. The Clone window opens.
4. Select the serial port you intend to use from the drop-down menu.
5. Select the correct Vendor and (if necessary) the appropriate Model.
6. Click OK to start the download process. (Smith, n.d.)

Connecting your Radio to CHIRP: Before you begin to use CHIRP with your radio, it is important to understand the two different modes of operation. Each radio falls into one of two categories: Clone and Live.

In Clone mode, the radio's memories are downloaded or uploaded all at once in a single clone operation. You can download an image of this type of radio, which includes all settings, including memories, VFO state, etc. The radio enters (or must be put into) a special mode of operation while communicating with the PC and is typically power-cycled after completion of a transfer.

In Live mode, the radio remains on and active during communication with the computer. Memories are transferred one at a time from the radio to the computer or from the computer to the radio. Changes to memories are made in real time against the radio as they are entered by the user (Smith, n.d.).

Now that you have an understanding of how to program your Baofeng radio with CHIRP, let's talk about import frequencies, repeater directories and customizing settings with CHIRP.

Importing Frequencies

1. Connect your Baofeng radio to your computer using a programming cable.
2. Download the latest version of CHIRP from the CHIRP website.
3. Open the CHIRP software, click on "Radio" then choose "Download from Radio".
4. The Port will be COM5, the vendor is Baofeng and the model will be listed in the drop-down menu.
5. Once your radio has completed the cloning process, a list will populate denoting all frequencies currently programmed into the radio.

6. From this menu, click on Radio>>Import from Data Source>>RepeaterBook.

7. Click on "RepeaterBook proximity query" to download the appropriate VHF/UHF channels for your area.

8. Import the desired frequencies into remaining open channels in your radio, (Michael, 2018).

Using Repeater Directories

1. Power on your Baofeng.

2. Hit the VFO/Memory button to enter "frequency mode" (VFO).

3. Hit the menu button, enter the number 7. You will want to hit the Menu button again and select OFF.

4. Now you type in your RX frequency.

5. Hit the menu button, enter the number 25. You will want to hit the Menu button again and select the desired +, - or OFF for your repeater offset.

6. Hit the Menu button, enter the number 26. You will want to hit the Menu button again and enter the actual offset of the repeater.

7. Hit the menu button, enter the number 13. This is where you will enter the PL tone/CTSS. (RadioReference 2013)

Customizing Settings with CHIRP

1. Open your radio image.

2. Click on the Settings Tab.

3. In CHIRP, click on the "Settings" tab on the left side and adjust the following:

a. Basic Settings
b. Carrier Squelch Level - 5
c. Beep - uncheck this box
d. Display Mode (A) - Name
e. Display Mode (B) - Name
f. Advanced Settings
g. Voice - OFF (Sherman, n.d.)

It's important to be aware that using this radio on specific frequencies that are restricted, without obtaining a license or approval from the FCC, could lead to several enforcement actions. These could include confiscation of equipment, imposition of fines, and other legal penalties. Therefore, it's crucial to always operate within the boundaries of the law.

We've taken the complexities out of setting up your radio, from entering frequencies by hand to using CHIRP. But remember, understanding the software is just the beginning. The *real* magic happens when you master the operations of your radio, transforming it from a mere device into a powerful tool for communication. In the next chapter, we are going to take a deep dive into the heart of your radio's functionality, and take the radio's features to the next level.

4

MASTERING BASIC OPERATIONS

"In a crisis, communication really matters. It must be embedded in strategy."

~ *Alistair Campbell*

Let me share the story of David, a client of mine and a community worker in New York, known for his proactive initiatives. I've known him for 5 years and have seen his passion for bringing people together. He told me he had once organized a city-wide clean-up drive. To coordinate the large number of volunteers, he purchased several Baofeng radios, underestimating the complexity of operating them.

David struggled with basic operations such as setting frequencies, saving channels, and adjusting the volume. On the day of the event, chaos ensued. Volunteers were scattered across the city, unable to communicate effectively due to the improperly set radios. The clean-up drive, which was supposed to bring the community together, ended up being a disorganized effort. This was all because of unfamiliarity with the basic operations of Baofeng radios. You wouldn't want to find yourself in a similar situation, would you?

The world of radio communication is filled with its own technical language—terms like "repeater", "CTCSS", "DCS", and "scanning"are frequently used. Initially, this might sound like technical jargon, but if you are interested in learning about your Baofeng radio, these words are essential tools that you need to master in order to significantly enhance your communication capabilities.

In this chapter, we will demystify these terms. We will learn the best ways to make and receive calls, understand how repeaters can help us communicate with more people, and look into the idea of privacy tones. We will also talk about how to effectively scan, to stay on top of busy bands and weather stations.

MAKING AND RECEIVING CALLS

Before you start making or receiving calls, ensure that your Baofeng radio is properly set up. This includes installing the battery, attaching the antenna, and turning on the radio. You should also be familiar with the basic controls such as the volume knob, channel selector, and PTT (Push-To-Talk) button. We have discussed these topics elaborately in the previous chapters. If you cannot recall, please go through Chapter Two again.

Making Calls

1. Select the Correct Frequency/Channel: Use the channel selector to choose the appropriate frequency or channel. Make sure you are licensed to operate on the selected frequency.
2. Listen Before Transmitting: Always listen for a few moments before transmitting to ensure the channel is clear.
3. Press the PTT Button: The PTT (Push-To-Talk) button is used to transmit. Only use this button to operate your radio with a ham license. Do not key up your radio without an antenna attached.
4. Speak Clearly and Slowly: When you're ready to speak, hold the radio about 2–3 inches from your mouth. Speak clearly and slowly into the microphone.
5. End Your Transmission: After you've finished speaking, release the PTT button to end your transmission. This allows others to respond.

Receiving Calls

1. Monitor the Channel: Keep your radio on and monitor the channel. You should hear other transmissions if you're on a busy channel.
2. Respond Appropriately: When you hear a call directed to you, press the PTT button to respond. Remember to listen before transmitting to avoid interrupting ongoing communications.
3. End the Conversation: When the conversation is over, sign off with your call sign to let others know that you're leaving the channel, (Nightingale, 2023).

UNDERSTANDING AND USING REPEATERS TO EXTEND YOUR COMMUNICATION RANGE

Repeaters are a crucial part of radio communication systems, including the Baofeng UV–5R, as they can significantly extend the communication range. A repeater is a device that receives a signal on one frequency and simultaneously retransmits it on another frequency. They are often located on high ground or tall structures, which allows them to cover a larger area.

When a signal is transmitted from a radio, such as the Baofeng UV–5R, it is received by the repeater. The repeater then retransmits the signal at a higher power and from a better location, allowing it to reach radios that would have been out of range of the original signal.

The use of repeaters can be particularly beneficial in areas where the terrain or buildings might otherwise block radio signals. By placing a repeater on a hill or tall building, it's possible to transmit signals over these obstacles.

The radio repeater operates by receiving a signal on one frequency, known as the input frequency, and simultaneously retransmitting it on a different frequency, referred to as the output frequency. The gap between these two frequencies is termed the offset.

To utilize a repeater, a transceiver capable of transmitting on the repeater's input frequency and receiving on its output frequency is required. Transceivers designed specifically for FM repeater operation are typically configured with the correct offset. Once the appropriate frequency is set, the repeater can be accessed by pressing the microphone button.

While most repeaters are open and accessible to anyone within range, some are restricted to specific groups, such as club members. These closed repeaters necessitate the transmission of a continuous sub-audible tone or a brief series of tones for access. These tones are known as Continuous Tone-Coded Squelch System (CTCSS) or Private Line (PL) tones, a term trademarked by Motorola.

Some repeaters, despite being open for public use, require special codes or sub-audible tones for access. The purpose of these access tones is to prevent inadvertent activation of the repeater due to unwanted transmissions.

There are several ways to find local repeaters:

- **Ask Local Amateurs:** Reach out to local amateur radio operators for information about local repeaters.
- **Contact Radio Clubs:** Local radio clubs often have information about repeaters in the area.
- **Use the ARRL Repeater Directory**: The ARRL publishes a comprehensive listing of repeaters throughout the United States, Canada, Central and South America, and the Caribbean.
- **Use TravelPlus for Repeaters™ CD-ROM:** This tool is handy for finding repeaters to use during vacations and business trips.

Using repeaters with your Baofeng UV–5R can significantly extend your communication range. However, always ensure you're licensed to operate on the frequencies you're using.

PRIVACY TONES

Privacy tones, also known as sub-audible tones, are used in two-way radio communications to reduce the annoyance of listening to other users on a shared channel. These tones are transmitted along with your voice when you press the Push-to-Talk (PTT) button. While *you* can't hear these tones, other radios that are programmed to listen for them *can*. This means that if two radios are programmed for the same privacy tone, they will only hear each other, and not any other radios on the same channel that are using a different tone.

There are two main types of privacy tones: Continuous Tone-Coded Squelch System (CTCSS) and Digital-Coded Squelch (DCS).

We have already talked a little bit about CTCSS and DCS in Chapter Three. They are a type of "privacy code" used in two-way radios, like your Baofeng, to reduce unwanted conversations on the same frequency.

You would use CTCSS and DCS when you want to communicate with a specific group of radios on a shared frequency, without being disturbed by other radios using the same frequency. For example, if you're part of a hiking group and all members are using Baofeng radios, you could all set the same CTCSS and DCS tone. This way, you would only hear transmissions from your group members, and not from any other people who might be using the same frequency.

CTCSS: To set up CTCSS on your Baofeng radio, follow these steps:

1. Press the key [MENU] to enter the menu.
2. Enter [1] [1] on the numeric keypad to get to receiver CTCSS.
3. Press [MENU] to select.
4. Enter the desired CTCSS sub-tone frequency in hertz on the numeric keypad.
5. Press [MENU] to confirm and save.
6. Enter [1] [3] on the numeric keypad to go to transmitter CTCSS.
7. Press [MENU] to select.
8. Enter the desired CTCSS sub-tone frequency in hertz on the numeric keypad. Make sure it is the same frequency as that you entered for receiver CTCSS.
9. Press [MENU] to confirm and save.
10. Press to [MENU] exit the menu system.

To turn CTCSS off, follow the same procedure but set it to off with the key instead of selecting a CTCSS sub-tone frequency, (Baofeng, n.d.).

DCS: To set up DCS on your Baofeng radio, follow these steps:

1. Press the key [MENU] to enter the menu.
2. Enter [1] [0] on the numeric keypad to get to the receiver DCS.
3. Press [MENU] to select.
4. Enter the desired DCS code on the numeric keypad.
5. Press [MENU] to confirm and save.
6. Enter [1] [2] on the numeric keypad to go to the transmitter DCS.
7. Press [MENU] to select.

8. Enter the desired DCS code on the numeric keypad. Make sure it is the same code as that you entered for receiver DCS.
9. Press [MENU] to confirm and save.
10. Press [MENU] exit the menu system.

To turn DCS off, follow the same procedure but set it to off with the key instead of selecting a DCS code, (Baofeng, n.d.).

Now, you may ask the question—What makes these two control methods different? While CTCSS and DCS are both privacy tones, they have their fair share of differences:

- **Signal Type:** CTCSS uses continuous tones below 300 Hz, whereas DCS uses digital data or encoded words.
- **Interference:** All encoded words in DCS can be used on the same channel without interference, which is not the case with CTCSS.
- **Bandwidth:** DCS signal spectrum occupies considerably more bandwidth than CTCSS1.
- **Development:** CTCSS was introduced by Motorola in the early 1960s, while DCS is considered an updated version of CTCSS.
- **Usage**: CTCSS is often used in amateur radio repeaters, while DCS is commonly used in professional, commercial, and amateur radio communications

Remember, CTCSS and DCS do not provide privacy. They do not prevent other radios from hearing your transmissions; they simply allow your radio to ignore unwanted signals.

SCANNING EFFECTIVELY WITH BAOFENG RADIO

Scanning is a crucial feature of the Baofeng radio. It allows you to monitor multiple frequencies and channels. Here are some steps to effectively scan with your Baofeng radio:

1. **Set the Radio to Frequency Mode:** Press the VFO/MR key.
2. **Set the Frequency Step**: Press the MENU key, then the 1/STEP key to view the current frequency step. To change the step, press MENU again, then use the up and down keys to select the desired frequency step. Once selected, press MENU again.
3. **Start Scanning:** Press and hold the */SCAN key for about two seconds. The display will show: "RANGE —:—."
4. **Enter the Frequency Range**: Enter the frequency range (in MHz) that you want to scan. The radio will begin to scan as soon as you enter the frequencies.

The Baofeng UV–5R is capable of scanning through channels and frequencies. Searching for active frequencies is a great way to discover new stations and keep track of activity in your area. To do this:

1. Set the Frequency Range: Enter the frequency range that you want to scan. For example, to listen for activity on the 2-meter amateur radio band, you would program the radio to scan between 144 and 148 MHz.
2. Set the Step Frequency: Depending on the band you're scanning, you'll need to set an appropriate

step frequency. For example, for the GMRS and FRS frequencies, program the radio with the frequencies 462:462 and a step frequency of 12.5 kHz.

Now that you have an understanding of scanning, you can try utilizing an advanced feature of your radio: Service Search. Service Search is a feature that allows the radio to scan through predefined frequency bands associated with certain services. For example, you might want to listen to local amateur radio repeaters, emergency services, or weather broadcasts. Unfortunately, the UV–5R does not have a built-in service search function like some other radios do. However, you can *manually* program the frequencies of interest and then scan through them. You can use websites like Radio Reference to find interesting local frequencies to program into your Baofeng radio.

Baofeng radios can also receive weather channels. For example, the Baofeng UV–9G has 30 GMRS pre-programmed channels & 11 NOAA channels with a weather radio scanning and receiving function. You can also find NOAA Weather and Emergency Frequencies and add those NOAA channels/frequencies into your UV–5R. These broadcasts are available 24/7 and provide the latest weather information, including severe weather alerts. The NOAA operates seven VHF frequencies ranging from 162.400 MHz to 162.550 MHz. You can program these frequencies into your UV–5R and then scan through them to receive weather updates.

By reading this chapter, you've mastered the basic operations of your Baofeng radio. You've learned how to make and receive calls, use repeaters to extend your communication range, understand privacy tones, and perform effective scanning. But this is just the beginning.

In the next chapter, we will unlock the advanced features of your Baofeng radio. We'll delve into the intricacies of using the dual watch feature, advanced menu settings, and also the special features that Baofeng radio has to offer.

5

UNLOCKING ADVANCED FEATURES

"Radio was supposed to die in 1945, when TV came along. It turns out that radio grew and grew, and it's a bigger business today than it has ever been."

~ *Alex Blumberg*

I remember talking to Maya, he resourceful lady that I met somewhere in the Midwest. She told me of a close call she had with fate, one particularly cold winter. As she was driving home, the weather quickly started to worsen. Luckily, weeks ago, Maya had programmed NOAA weather alerts onto her Baofeng, a seemingly unnecessary precaution at the time. Now, it was her lifeline.

She received an alert that there was a blizzard coming her way – she would not be able to make it home in time. Thus, she decided to immediately seek shelter in a nearby motel, and avoided potentially getting trapped in the snow.

You might feel that you have already learned *all* there is to know about your radio. But in reality, there is so much more to it. Understanding advanced features, like Dual Watch function, VOX setup, and dual standby mode is going to help you unlock the full potential of your device. These features will be particularly useful in emergency situations like natural disasters, search and rescue operations, power outages, and so on.

Their effectiveness depends on the skill and knowledge of the user. That's why it's important to familiarize yourself with these features before you find yourself in such an emergency. In this chapter, we will delve into these advanced features, exploring the depths of what your Baofeng radio can do. From the dual watch function to the bandwidth selection, VOX setup, and TDR mode, we will learn how to transform your pocket gizmo into a reliable partner that you can count on when it matters the most.

DUAL WATCH FUNCTION

The Dual Watch function is a feature that allows the radio to monitor two different frequencies or channels at the same time. This function is also known as TDD (Time-Division Duplex). The Baofeng UV–5R radio has a single built-in receiver but can "watch" two channels (semi-duplex). It can monitor two different frequencies, even on different bands (VHF/UHF), and the radio will monitor *both* frequencies, giving priority to the first station to receive an incoming call. (Baofeng Manual, n.d.)

In practical situations, the Dual Watch function can be incredibly useful. For instance, if you are in a situation where you need to monitor two different frequencies or channels simultaneously, such as during a hiking trip or a camping adventure where you might need to stay connected with two separate groups, the Dual Watch function comes in handy. You can keep track of communications on both channels, ensuring you don't miss any important updates or calls.

While the Dual Watch function allows you to monitor two frequencies, it's important to note that the Baofeng UV–5R only has a single receiver and a single speaker. This means that while it can scan both frequencies, it can only receive and play audio from one frequency at a time.

Here is how you can use the Dual Watch function in your Baofeng radio UV–5R:

1. **Activating Dual Watch:** To activate the Dual Watch function, you need to go to the menu, select option 7 (TDR), press the Menu button again, press the up-arrow button to turn it on, and finally press the exit button.
2. **Using Dual Watch:** Once the Dual Watch function is activated, the radio will start scanning both A and B frequencies. The radio will give priority to the first station to receive an incoming call. This means if a signal comes on A first, then it will lock on and ignore B.
3. **Switching Between Frequencies:** You can toggle between A and B by pressing the "Enter" button. This allows you to select which frequency to transmit on. Remember, this is dual watch, and not dual receive.
4. **Handling TDR and Scan Mode Simultaneously**: If you want to scan and handle TDR mode at the same time, you can remove the frequency of the other VFO from your

scan frequencies, activate Dual Watch (TDR) mode, activate scan, and have a friend transmit on the frequency of the other VFO, (Stack Exchange, 2021).

ADVANCED MENU SETTINGS

Bandwidth Selection: When it comes to bandwidth selection, Baofeng radios typically have two options: Narrow and Wide. These terms refer to the width of the radio channel. A wide channel is 25 kHz, while a narrow channel is 12.5 kHz.

The choice between narrow and wide bandwidth depends on a few factors:

- **Regulations:** Some frequencies or applications may require the use of narrowband to avoid interference with other signals.
- **Sound Quality:** Wideband can provide better sound quality, but it uses more power and has a higher chance of interference with other signals.
- **Range**: Some users have found that setting the radio to narrowband can improve the range of the radio.

It's important to note that if you set your radio to receive in narrowband and the transmission is in wideband, it may distort the transmission. Conversely, if you're transmitting in narrowband and someone is listening in wideband, they might not notice any difference.

VOX Setup: VOX stands for Voice Operated Switch. It's

a feature that allows you to transmit automatically without needing to press the transmit button. This can be particularly useful in situations where you need to have your hands free, but it's important to use it properly to avoid causing confusion on radio nets.

Here's how you can set up the VOX feature on your Baofeng radio:

1. Access the VOX Function: You'll need to find the VOX function from the Menu on your Baofeng radio.

2. Set VOX Sensitivity: Once you've accessed the VOX function, you can select the desired level of sensitivity for your environment. The highest number represents the most sensitive level for very quiet locations.

3. Activate VOX Setting: After setting the VOX sensitivity, push the Push-to-Talk key or Menu option to go to the next feature. Your VOX setting will now be activated, (LiGo, 2021).

It's important to note that the VOX feature can pick up any noise in the room and turn on the transmitter. Therefore, it's best to use this feature in a quiet room to avoid unwanted transmissions.

Also, bear in mind that radios aren't like phones or voice chat. If you're transmitting, you cannot hear others' transmissions, and if more than one person is transmitting on the same frequency (aka channel), the strongest signal (usually whichever radio's closest to that particular receiver) wins, reaching the radios that are receiving.

SPECIAL FEATURES OF A BAOFENG RADIO

Baofeng radios have a number of special features that really improve the overall experience. Getting to know them will help you enjoy this powerful tool even more than you already do.

FM Radio: The Baofeng radio, particularly the UV–5R model, is capable of receiving FM radio. This means you can tune into your favorite FM radio stations using this device. The reception quality is quite impressive, with users reporting the ability to listen to radio stations located approximately 50 miles away. This makes the Baofeng radio a great companion for outdoor adventures where you might want to catch up on news, music, or weather updates.

Accessing the FM radio feature on the Baofeng radio is straightforward. This feature adds to the overall utility of the device, making it far more than just a communication tool. Here is how you can use the Baofeng radio to receive FM radio:

1. **Enter Frequency Mode**: Press the [VFO/MR] button to enter Frequency Mode.
2. **Choose the A Side**: Press the [A/B] button and choose the A Side (upper display).
3. **Select the Frequency Band:** Press the [BAND] button for the frequency band.
4. **Disable TDR** (Dual Watch/Dual Standby).
5. **Enter the Frequency:** Now, you can enter the frequency of the FM radio station you want to listen to.

6. *Optional* - Enter the Transmit CTCSS/DCS
   Code: If necessary, you can enter the transmit
   CTCSS/DCS code, (Morgan, 2023).

While the FM radio is being activated, press the I*SCAN
key to search for FM radio stations. This will allow the radio
to automatically scan through the available FM radio
frequencies until it finds a station.

The FM radio feature of the Baofeng radio enhances its
appeal by providing users with the ability to stay connected
with the world through music, news, and more. It's a feature
that truly makes the Baofeng radio a versatile and valuable
gadget to have.

Alarm Function: The Baofeng radio, particularly the
UV–5R model, is equipped with an alarm function that can
be activated or deactivated by pressing and holding the
CALL button. This feature can be useful in emergency
situations where you need to draw attention or signal for
help.

The alarm function has three settings:

- SITE: Local, just comes out the radio.
- TONE: Transmits tone.
- CODE: Transmits DTMF code, ("Baofeng UV–
  5R Manual", n.d.).

If the alarm is set to TONE or CODE, it's going to transmit
on whatever frequency the radio is set to. So, if you are on a
repeater and hit it, everyone listening to that repeater is
going to hear your radio making annoying sounds and you
are in effect jamming the repeater. It's not something you

want to be doing and if you do it enough you will get caught by other ham radio operators for causing interference on a repeater - in radio-land this is known as being "fox-hunted".

In addition to the standard alarm function, the Baofeng radio also has a unique feature known as the "FM radio alarm mode." This mode allows the radio to automatically switch to the FM radio when the alarm is activated. This can be particularly useful in situations where you need to stay updated with news or weather updates during an emergency.

To activate the alarm function, you simply need to press and hold the CALL button. To deactivate it, you press and hold the CALL button again. It's a simple yet effective feature that adds to the overall functionality of the Baofeng radio.

Voice Prompts: The UV–5R comes with a plethora of features; one I really like is Voice Prompts. This feature can be particularly useful for those who are new to the device, or for situations where you need to operate the radio without looking at the screen. Here's how you can use it:

1. Power on the radio. The antenna does not need to be connected.
2. Press the MENU button on the keypad. While programming, place your radio on LOW power. To do this, press the # button. You will see an "H," "M," or "L" in the top-left corner of the screen, indicating HIGH, MEDIUM, and LOW power.
3. The UP and DOWN arrows allow you to navigate through all menu options.
4. You can jump to specific options by typing in that option's menu number. For example, pressing

MENU and keying in "14" will bring you to option 14, which controls the radio's voice prompts.

5. Once you navigate to the option you wish to modify, press the MENU button a second time to access the option's available choices.

6. Select the choice you want to use with the UP/DOWN arrows, or by typing in the necessary value with the keypad.

7. Then press the MENU button a third time to SAVE/CONFIRM the choice. (Olander 2022)

If you're tired of the UV–5R giving you annoying voice prompts, you can turn this option off anytime. However, these voice prompts can be helpful when you're learning to navigate the radio's menu or when you're in a situation where you can't look at the screen.

With the conclusion of this section, you've successfully mastered the advanced features of your Baofeng radio. You've learned how to use the Dual Watch function, navigate through advanced menu settings, and even discovered some special features like the FM radio and alarm functions.

As you can see, the world of Baofeng radio is vast and there's always more to learn. In the next chapter, we're going to look at how you can optimize your signal with antennas. This will take your Baofeng experience to a whole new level.

6
OPTIMIZING YOUR SIGNAL WITH ANTENNAS

"TV gives everyone an image, but radio gives birth to a million images in a million brains."

~ *Peggy Noonan*

In my years of experience with Baofeng radios, I've learned that the antenna is not just an accessory, but the heart of any radio system. It's the bridge that connects us to the vast realm of radio waves. It is the magic wand that will make your voice travel across terrific distances, and the sensitive ear that can pick up faint whispers from far-off locations.

Whether you're a beginner just starting out with your first Baofeng radio, or a seasoned ham radio operator, this chapter will provide valuable insights into the science and art of antennas. We will start by unraveling the basics of radio antennas, including their size, type, and frequency. We will then move on to a comparative analysis of popular aftermarket antennas for Baofeng radios, helping you choose the one that best suits your needs.

By the end of this chapter, you will not only have a deeper understanding of antennas but also the confidence to optimize your Baofeng radio signal for the best possible performance.

BASICS OF RADIO ANTENNAS

Radio antennas are essential components of Baofeng radios. The primary function of an antenna is to radiate or receive radio waves. In the context of a Baofeng radio, the antenna is responsible for transmitting and receiving signals over various frequencies. The standard antenna that comes with a Baofeng radio is typically built for 136-174MHz and 400-480MHz amateur radio bands.

The size of an antenna is directly related to the wavelength of the frequency to be received. This is because antennas must maintain a certain size in order to work efficiently with electromagnetic radiation. For instance, an antenna designed to transmit and detect green light would need to be approximately 250 nm in length (Stone et al. 2017).

When comparing antenna sizes, those at lower frequencies (like Bluetooth 2.4 GHz) are usually much bigger than those at higher frequencies (like 5G 28 GHz). This size discrepancy arises due to the wavelength differences between low and high frequencies, where lower frequencies necessitate larger antennas for optimal performance, while higher frequencies enable the design of more compact antennas.

An antenna's performance also depends on its type. Depending on direction, there are mainly three different types of antenna:

- **Omni-directional Antenna:** This kind of antenna emits radio power equally in all directions that are perpendicular to an axis (azimuthal directions). The power varies with the angle to the axis (elevation angle), diminishing to zero on the axis itself. These antennas are commonly used in radio broadcasting and in mobile devices that utilize radio waves, such as cell phones, FM radios, walkie-talkies, wireless computer networks, cordless phones, GPS devices, and base stations that communicate with mobile radios. There are several types of low-gain omnidirectional antennas, including the whip antenna, "Rubber Ducky" antenna, ground plane antenna, vertically oriented dipole antenna, discone antenna, mast radiator, horizontal loop antenna, and halo antenna. (Commscope, n.d.)

- **Semi-directional Antenna:** Semi-directional antennas concentrate their energy output within a limited area. They are typically employed in scenarios where one antenna must interact with multiple others within a designated zone, often referred to as point-to-multipoint situations. These antennas are engineered to guide the RF signal towards a specific direction, facilitating point-to-point communication. Semi-directional antennas are commonly used for indoor or outdoor communication spanning short to medium distances. (Access Agility 2018)

- **Directional Antenna**: Also known as a beam antenna, a directional antenna is designed to transmit or receive a higher degree of radio wave power in certain directions. The Yagi antenna, the log-periodic antenna, and the corner reflector antenna are among the most prevalent types of directional antennas. These antennas are frequently used when there's no need for a wide coverage area. They have the ability to amplify the power transmitted to receivers in their direction, or minimize interference from undesired sources, (Access Agility, 2018).

While we usually prefer omni-directional antennas for our Baofeng radios, we can replace them with directional or semi-directional antennas based on our needs.

The best antenna for a particular situation depends on a variety of factors, including the frequency of operation, the physical constraints of the device it's attached to, and the desired directionality and gain of the antenna. The size and type of an antenna are crucial factors when it comes to determining its efficiency. That's why understanding the characteristics of different forms of antennas is going to help you select an antenna for your specific device.

POPULAR AFTERMARKET ANTENNAS FOR BAOFENG RADIOS

Aftermarket antennas are replacement antennas that you can purchase to replace or upgrade the original antenna that came with your device. These antennas are often used to improve the performance of your device, such as enhancing signal reception or transmission. They can be particularly useful if the original antenna is damaged, not performing well, or if you simply want to customize your device.

There are multiple aftermarket antennas available for Baofeng radios that can help to enhance performance or efficiency of the popular two-way radio:

Nagoya NA–701C: The Nagoya NA–701C is a high-performance antenna designed for use with Baofeng radios. This antenna is specifically tuned for transmitting and receiving on the commercial bands, optimized to 155/455 Mhz. This black 8-inch antenna has a very sleek design that is going to make your Baofeng radios look even better.

The Nagoya NA–701C was specifically designed for the higher frequencies commonly used on commercial, government, and GMRS frequencies. It is said to receive and transmit signals over long distances as far as 25,000 meters. However, since it is matched specifically for 155/455 MHz, it may not perform as well outside of this range. Another downside to this radio is its long antenna which might not be very convenient for portable use, (Baofeng Tech, n.d.).

ABBREE AR–152A12: The ABBREE AR–152A12 is a foldable, military-style antenna that extends to a length of 18.89 inches. It has an effective antenna length of 15.24 inches (5/8 wave) and, according to the manufacturer's specifications, should provide a 3dB gain over a typical dipole antenna. This antenna is compatible with any radio that uses an SMA female connection, and that includes most of the popular Baofeng radios.

The ABBREE AR–152A12 antenna is specifically tuned for 462–467 MHz, making it ideal for GMRS radios. However, just like the Nagoya NA–701C, it may not perform as well outside of this range.

The ABBREE AR–152A12 antenna has been reported to significantly increase the range of handheld radios. While the exact range can vary based on the specific radio model and the environment in which it's used, it has been tested to hit a repeater 28 miles away, across a developed urban area, (The Beaten Path, n.d.).

The Radtel Foldable Tactical Antenna: This is a full band walkie talkie antenna that operates in the frequency range of 136–520MHz. It is designed to boost the receiver range of your radios, making it an excellent choice for extending the signal of walkie talkies during outdoor activities such as hunting, hiking, traveling, and camping. This antenna also has an SMA–Female connector and is thus compatible with most Baofeng radios.

One of the standout features of the Radtel Foldable Tactical Antenna is its portability. It can be folded into two or three folds, making it light and easy to carry. You can easily put it in a pocket or bag. This feature, coupled with its ability to boost signals, makes it the perfect companion for outdoor adventures.

This antenna is made from non-deformable rubber materials and high-gain antenna heads, ensuring long-term use and stable quality. It has a gain of 2.15dBi, a maximum power of 8 Watts, and an impedance of 50 ohms. The antenna length varies, with options of 33cm, 47cm, 72cm, 108cm, and 124cm available, (Radtels, 2023).

The Baofeng Magnetic Antenna: The Baofeng

Magnetic Antenna is a high-performance antenna that is particularly useful for vehicle-based operations, making it an excellent choice for those who need to use their radios on the go. It operates in the frequency range of 136–174 MHz/ 400–520 MHz.

One of the key features of the Baofeng Magnetic Antenna is its magnetic base. This allows for easy mounting on vehicles, ensuring a stable and secure fit even when the vehicle is in motion. This feature, combined with its ability to boost signals, makes it an ideal choice for those who frequently use their radios in a vehicle.

Here is a detailed comparison between the four popular aftermarket antennas:

Antenna	Frequency Range	Gain	Max Power	VSWR	Impedance	Length
Nagoya NA-701C1	155/455MHz	2.15 dBi	10 Watts	Less 1.5:1	50 OHM	21 cm (~8 inches)
ABBREE AR-152A12	144/430MHz	2.15 dBi	10 Watts	Less 1.5:1	50 OHM	48cm
Radtel Foldable Tactical Antenna	136/520MHz	2.15 dBi	8 Watts	Less 1.5:1	50 OHM	33cm/47cm/72cm /108cm/124cm
Baofeng Magnetic Antenna	136/174 MHz / 400/520 MHz	3 dBi	8 Watts	Not specified	Not specified	12cm

Please note that the length of the Radtel Foldable Tactical Antenna can vary depending on the model. Also, the performance of these antennas can be influenced by other factors such as the specific radio model they are used with, the environment, and the height above ground. That's why it's recommended to choose the antenna based on your specific needs and usage scenarios.

DIY ANTENNA PROJECTS

DIY (Do-It-Yourself) antenna projects are a popular aspect of the amateur radio hobby. They involve creating your own antennas for transmitting and receiving signals. The type of antenna you choose to build depends on your specific needs and the frequency bands you plan to operate on.

In most cases, these DIY antennas are cheaper than pre-made ones. The cost of a DIY antenna largely depends on the materials you use, many of which you might already have at home. For example, you can make a simple antenna with items like wire, a piece of wood, and some basic tools. On the other hand, buying a new antenna can cost anywhere from $30 to $300, (that is not including the cost of installation.)

For radio enthusiasts, it's not just about saving money; it's about the valuable learning experience. Hands-on practice' is a great way to learn about the complicated aspects of radio wave transmission and antenna theory. You can also customize your antennas any way you like. By building your antenna, you gain the flexibility to tailor it to your specific requirements. This could involve optimizing it for a particular frequency band or configuring it to be highly directional, focusing on a specific source.

It's impossible to overstate how *satisfying* it is to use tools that you helped *make*. Besides giving your radio setup a unique look, this sense of success makes the whole experience worth it. The significance of a DIY antenna lies in its potential to enhance your radio's performance. A well-constructed DIY antenna can boost the range and clarity of reception, proving particularly beneficial in locations with weak signal strength, or when tuning into specific frequencies not optimized by your current antenna.

While there are clear advantages, it's essential to weigh the pros and cons of building a DIY antenna. On the positive side, it is cost-effective, educational, and customizable. However, it demands time and technical knowledge. The construction process may be time-consuming, especially for beginners. It also requires a certain level of technical understanding to be able to build an antenna from scratch, nevertheless, here is the basic step-by-step guide on how to build a simple DIY antenna for your Baofeng radio:

1. **Step 1:** Choose the Frequency: Decide on the frequency you want to receive. This is important because it determines the length of your antenna.
2. **Step 2:** Get a Conductor: Choose a good conductor. For example, you can use a copper wire and a piece of copper plating.
3. **Step 3:** Determine the Wire Length: Calculate the length of the wire you need. This is done by dividing the speed of light (c = 299,792,458 m/s) by the frequency (f). For example, if you're building an antenna for 2.45GHz, the wavelength is 12.236 cm. Usually, monopole antennas are quarter-wavelength structures, so the length of the wire should be 12.236/4 = 3.0509cm.
4. **Step 4:** Cut the Wire: Cut the wire to the length you calculated in the previous step.
5. **Step 5:** Strip the Insulation: Strip the insulation off the end of a coaxial cable to create a simple antenna. Get a coaxial cable that's long enough to go from your TV to the nearest window so you can get the best reception.

6. **Step 6:** Attach the Antenna to Your Radio: Run the other end of the cord to the port on your radio to attach the antenna.

7. **Step 7:** Test the Antenna: Once your antenna is in place, check the settings on your radio to make sure the input is set to "Antenna" or "Air" instead of "Cable," (Crider, 2019).

Building a DIY antenna for your Baofeng radio is a straightforward process. This process not only enhances your understanding of radio frequencies and antenna design, but also provides a cost-effective solution for improving your radio's reception. But remember, safety should be your priority when working on DIY antenna projects—always make sure your antenna is properly installed and grounded to avoid any potential hazards.

THE SCIENCE OF RADIO WAVES

We have talked a lot about Ham Radio, and radio signals. But do we truly know how radio waves travel? I have met many radio enthusiasts who know a lot about radios but without knowledge of the *physics* of radio waves. But this book aims to teach you everything about Ham Radios, including how they work the way they do.

Radio waves, just like light waves, microwaves, and X-rays, are a type of electromagnetic wave. However, radio waves have the longest wavelengths in the electromagnetic spectrum. They range from the length of a football, to larger than our planet!

Radio waves are generated by charged particles undergoing acceleration, such as time-varying electric currents. These waves are naturally emitted by phenomena like lightning and celestial bodies, and are a component of the radiation given off by all objects that have heat. Man-made radio waves have a wide range of uses, including radio communication, radar, computer networks, broadcasting, and various navigation systems.

The journey of radio waves involves numerous electromagnetic events, including reflection, polarization, refraction, absorption, and diffraction. The characteristics of radio wave propagation in the Earth's atmosphere vary with frequency. Long waves can bend around obstacles, adapting to the landscape, while short waves bounce off the ionosphere and return to Earth beyond the horizon. However, the travel distance of both types of waves is restricted to the visual horizon, as short wavelengths bend, or diffract minimally, and follow a line of sight. (Britannica, n.d.)

Now, this all might seem like a load of technical jargon and you must be wondering how YOUR radio signals travel far distances? Radio signals from your Ham Radio travel through the air in the form of electromagnetic waves. Here's a simplified explanation of how it works:

- **Transmitting the Signal:** When you speak into your Ham Radio, your voice is converted into an electrical signal. This signal causes electrons in the radio's transmitter to oscillate.
- **Creating Radio Waves:** These oscillating electrons rush up and down the transmitter, creating radio waves. The frequency of these waves depends on the rate of oscillation. Ham Radios can use many frequency bands across the radio spectrum.

- **Traveling Through the Air:** The radio waves then travel through the air at the speed of light. Higher frequencies travel with more energy, but cover a much smaller distance, while lower frequencies require less energy to travel much longer distances.
- **Receiving the Signal:** When these radio waves hit a receiver (like another Ham Radio), they make the electrons inside it vibrate. This vibration recreates the original signal.
- **Converting Back to Sound:** The receiver then converts these vibrations back into sound waves, which you hear as the original voice message, (Woodford *et al.* 2022).

It's important to note that radio waves normally travel in straight lines. However, they can also reflect off objects, refract around objects, and diffract around edges and corners. This is why you can receive radio signals even if you're not in the direct line of sight of the transmitter.

We've now learned a lot about radio antennas and their science. While antennas are a great way to improve your radio's performance, it's important to learn how to keep your antenna at peak performance, ensuring that every signal sent and received is as clear, and strong, as possible.

In the next chapter, we'll talk about maintaining and preserving the utility of our Ham Radio and making sure we can enjoy its exciting features for a long time.

7

KEEPING YOUR BAOFENG AT PEAK PERFORMANCE

"Radio is the most intimate and socially personal medium in the world."

$\sim$ *Harry von Zell*

Nothing lasts forever, not even your highly durable Baofeng radio. It's performance is highly correlated with how you take care of it. There might be times when you find yourself in situations where your pocket radio starts malfunctioning. What do you do then? And how do you fix it?

My friend Karl, a storm chaser, considers his Baofeng radio an indispensable ally. Once however, during a lightning storm, his radio abruptly fell silent, severing his connection with his team just as a tornado spiraled down from the clouds. In a state of panic, he scrambled for her spare batteries, only to discover them mysteriously depleted.

We later spoke about his experience and I recognized his batteries had likely suffered from the "phantom drain" phenomenon, a stealthy power leech that slowly siphons energy from radios even when they appear to be off. I advised him to invest in a high-quality battery case equipped with an on/off switch.

Our devices are not immune to problems, and given our radios' essential role in keeping us prepared and safe, these issues can become life-threatening. It's important that we have a good understanding of these common problems so that we can fix them when needed.

We have already learned a lot about the functionality of Baofeng radios. However, when it comes to troubleshooting, we still have some ground to cover.

In this chapter, we'll discuss common user issues and provide step-by-step solutions. I will guide you through a maintenance schedule, advise on when to clean contacts, replace parts, and more. We'll also talk about firmware updates and long-term care, including tips on battery life cycle, storage, and protection against moisture.

COMMON USER ISSUES—AND THEIR SOLUTIONS

While Baofeng radios are very rugged and durable, like any other electronic device, they do sometimes experience issues.

Slow Scan Rate Issue: Baofeng radios are known for having an issue with slow scan rate. This means that when the radio is set to scan mode, it takes a considerable amount of time to cycle through the frequencies or channels. This can be problematic because by the time the radio scans back to a particular frequency, the conversation or transmission on that frequency might already be over.

Potential Solution:

- **Determine the Frequencies**: First, determine what frequencies are in use in your area and then program these into the memory of your UV–5R. There are many resources out there that can tell you which channels are being used.
- **Use Multiple Radios**: If you have multiple radios, set each one up as an individual "bank" of memory channels. For example, one radio for FRS/GMRS, one radio for Public Safety, one radio for 2 meter, etc. This method helps to compensate for the sluggish scan speed.
- **Use an RTL–SDR Dongle**: If you have an RTL–SDR dongle, you can use it to find the stations so you can then program them into the UV–5R. The RTL–SDR software, SDR# along with a free scanning suite called "Frequency Manager Suite" will find and save-to-file all the signals it finds using a predetermined scanning block of frequencies. This scans much faster compared to the UV–5R.
- **Change Scan Resume Setting**: If your radio locks onto a channel while scanning and remains on that channel for several seconds before resuming scanning, you can change the scan resume setting to CO (carrier operated) instead of TO (timeout). In Chirp, this setting is under Settings/Advanced Settings/Scan Resume. On the radio, it's MENU 18 SC-REV.

Weird Display Issue: This issue is characterized by the LCD controller defaulting to a certain state when powered up. The top line of the display might appear dark, while the bottom line is light. This could be due to a problem with the processor, or the connection between the processor and the screen. If the radio still works, the processor might not be dead, but something could be preventing it from functioning properly.

Potential Solutions:

- Try to get a response out of the radio with the programming cable. If it responds, try loading a default config. (This is a bit of a long shot, but it might be worth a try.)
- If you're comfortable with electronics, you could try taking the radio apart and poking around with a multimeter. This might help you identify what's wrong.
- If all else fails, you could remove the CPU and interface directly to it. This would allow you to potentially repurpose the radio.

Radio Beeps but Won't Transmit: If your Baofeng radio isn't transmitting, but simply beeps, even though the LCD works and it can receive signals, the issue might be that the battery supply's voltage is too high. The radio has a safety built-in. This prevents transmission when voltage is above 8V. This measure is to keep the transmitter from burning out.

Potential Solution:

- Check the battery voltage and replace it if necessary.

FIRMWARE UPDATES

If you are not a tech savvy person, you might not be familiar with this term. Firmware is essentially the "software for hardware" and is typically embedded in a piece of hardware. You can think of it as the instructions that tell your hardware how to function.

Firmware is stored in non-volatile memory devices such as ROM, EPROM, EEPROM, and flash memory. This means that the firmware remains intact even when the device is turned off. It's not something that can be easily changed, or deleted, by the user.

Examples of devices that contain firmware are: computers, smartphones, cameras, and even home appliances like microwave ovens and remote controls. The firmware in these devices controls how they function and interact with other devices, (Fisher, 2023). And just like these devices, your Baofeng radio has its own firmware.

Since firmware is basically the software for your radios, it can be updated from time to time. However, these updates will come with their fair share of benefits and risks.

The reason why most people update their devices is because of the additional features they provide on each new update. But the benefits of firmware updates extend beyond the introduction of new features. They breathe life into the performance of devices, making them more efficient and reliable. Whether it's the speed of APRS data downloads, or the quality of Bluetooth connectivity, firmware updates fine-tune these aspects, ensuring your device performs at its best.

Firmware updates also bring about improvements in the user interface, making it more intuitive and user-friendly. From minor text corrections for improved readability to the addition of new menus for easier navigation, each update refines the user experience.

One of the most significant benefits of firmware updates is the optimization of battery life. Some updates introduce features like a battery save scanning mode, which allows for better battery optimization during long-term scanning sessions.

However, updating the firmware of your Baofeng radio comes with certain risks that users should be aware of. One such risk is the potential for errors if the update is not done correctly. For example, a user reported losing the lettering on the screen of their Baofeng DM–1701 after a firmware update. In addition, users have reported that if these devices are not properly programmed within their approved frequency ranges, they may inadvertently transmit on restricted frequencies after a firmware update. This can interfere with critical communications, jeopardizing public safety and violating regulatory requirements.

So, while firmware updates are a breath of fresh air for your device, they do come with challenges. It's up to you to decide if you wish to take the risk or not.

Now, before we learn how to update our firmware, it's important you know what version your Baofeng radio is running. Here is how you can find out your firmware version:

1. **Turn off the Radio:** Make sure your Baofeng radio is turned off before you begin.

2. **Hold the Number 3 Key:** While the radio is
   off, hold down the number 3 key on the keypad.
3. **Power on the Radio:** While still holding the
   number 3 key, turn on the radio.
4. **Check the Display:** The firmware version
   should now display on the screen of the radio.

Please note that this method may vary depending on the
model of your Baofeng radio. Always refer to the specific
user manual for your radio model.

The process of updating your firmware is pretty
straightforward. It is pretty similar to updating the software
on your computer. Here are the steps:

1. **Check the Current Firmware Version:**
   First, you need to check the current firmware
   version of your Baofeng radio.
2. **Visit the Baofeng Website:** Go to the official
   Baofeng website or the website of the radio's
   manufacturer. Look for a section on firmware
   updates or downloads.
3. **Download the Latest Firmware:** If a newer
   firmware version is available, download it.
4. **Connect Your Radio to Your Computer:**
   Using a USB programming cable, connect your
   Baofeng radio to your computer.
5. **Launch the Programming Software:**
   Open the programming software on your
   computer. If you don't have it, you can download it
   from the Baofeng website as well.
6. **Update the Firmware:** Follow the
   instructions provided by the programming
   software to update the firmware on your radio.

7. **Reboot Your Radio:** After the update is complete, reboot your radio to ensure the new firmware is loaded, (Baofeng Manual, 2023).

If you're not comfortable performing the update yourself, or if you're unsure, you can always find help in the manual of your specific radio device.

CARING FOR YOUR BAOFENG RADIO

Baofeng radios are all very sturdy devices. If you can take care of them, they can last a lifetime. However, to ensure their longevity and optimal performance, it's crucial to understand how to care for these devices in the long term. I have seen many individuals who have neglected these best practices and complained about the longevity of these radios. I sincerely believe taking good care of your radio will extend the life of your Baofeng radio and maintain its performance over time. Here is how:

Battery Life Cycle

Baofeng radios typically come with a Lithium-ion battery. The battery life of a Baofeng radio can last up to 12 hours, and even a few weeks for some models. However, the actual battery life can vary depending on various factors. Some of the factors that influence your battery performance include:

- **Usage**: Obviously, the more you use the radio, the faster the battery will drain. This includes both transmitting and receiving. Transmitting, especially at high power, uses more energy than receiving.

- **Settings:** Certain settings can impact battery life. For example, the backlight on the screen, if left on, can drain the battery faster. The volume level can also affect battery life.
- **Environmental conditions:** Extreme temperatures, both hot and cold, can affect battery performance and life, (Baofeng Manual, n.d.).

Most of these factors are basically unavoidable. So instead of trying to influence the factors you can incorporate some basic practices that are going to help you prolong the life of your battery:

- **Charge in normal room temperatures:** Lithium-ion batteries prefer moderate temperatures for charging. Extreme cold or heat can reduce the battery's efficiency.
- **Turn off the radio for a faster charge**: When the radio is off, all the charge from the charger goes to the battery, which can speed up the charging process.
- **Don't interrupt the charging process:** Unplugging the power to the charger or removing the battery and/or radio before it's finished charging can affect the battery's capacity over time.
- **Storage of Battery:** If you're not going to use the radio for a while, it's recommended to store the battery separately with about 40% charge. This can help maintain the battery's health during storage, (Baofeng Manual, n.d.).

Storage Tips

Proper storage of your Baofeng radio is crucial for maintaining its longevity and performance. The radio and its components, especially the battery, are sensitive to environmental conditions and improper storage can lead to decreased performance or even damage.

When storing your Baofeng radio, there are a few key points to keep in mind:

- **Battery Charge:** Store the radio with the battery at around 40% charge. Storing a Lithium-ion battery at 100% charge is less than optimal, and storing it in a discharged state can permanently damage it.
- **Temperature:** Lower temperatures are better for storage, but avoid freezing temperatures. Extreme temperatures can affect the battery's efficiency and lifespan.
- **Separate Storage:** Leaving the battery connected to the radio or the charger might discharge it slowly due to leakage current. Therefore, it's recommended to store the radio and battery separately.
- **Avoid Sunlight and Heat:** Avoid any place in the sun or that gets hot, like a car. Exposure to direct sunlight or high temperatures can cause damage to the radio and battery, (Amateur Radio Stack Exchange, 2015).

It is important to note that each Baofeng radio model may have specific storage recommendations, so always refer to the user manual for model-specific instructions.

Protection against Moisture

While some Baofeng radios are designed with water-resistant materials, not all models are waterproof. Even a small amount of moisture can cause significant damage to your device and reduce its performance. So, it's always better to be safe than sorry. Here are some detailed tips to prevent moisture damage:

- **Avoid Exposure to Moisture:** This seems pretty obvious but so many get it wrong. Try to keep your radio dry at all times. Avoid exposing it to rain or other sources of moisture. If you're in a moist environment, consider using a waterproof case or bag to protect your radio.
- **Dry Thoroughly If Wet:** If your radio does get wet, it's important to dry it thoroughly before using it. Remove the battery and any other removable parts, then gently wipe down the surfaces of the radio with a dry cloth. You can also shake the radio to remove any water from the crevices.
- **Do Not Submerge Unless Waterproof:** Never submerge your radio in water unless it's specifically designed to be waterproof. Some Baofeng models, like the UV–9G, are designed to be waterproof and can be submerged in up to 1 meter of water for up to 30 minutes. However, even these models should not be left in water for extended periods.
- **Store in a Dry Place:** When not in use, store your radio in a dry place. This helps to prevent any moisture in the air from getting into the radio.

Avoid storing it in areas with high humidity or risk of water exposure.

- **Use Desiccants:** If you're storing your radio for a long period, consider using desiccants, like silica gel packets, to absorb any moisture in the storage area, (Baofeng Manual, n.d.).

Keeping your Baofeng radio at peak performance is a matter of understanding common user issues and their solutions, staying updated with firmware, and practicing long-term care. The longevity of your radio is not just about how you *use* it, but also how you *care* for it. Your Baofeng radio is a robust piece of technology, and with the right care, it can serve you well for many years to come.

In the next chapter, "The Baofeng in Action: Case Studies and Scenarios," we will explore real-world applications and scenarios of Baofeng radios. You'll see how all the knowledge you've gained so far comes into play, in practical situations.

8

THE BAOFENG IN ACTION: CASE STUDIES AND SCENARIOS

"Radio gives the listener a platform to dream."

~ David Bowie

In my lifetime, I have owned my fair share of devices, but none have left an impression quite like the Baofeng radio. I remember a particular incident during a mountaineering expedition. We were caught in a sudden snowstorm, and our cell phones had lost all connection.

It was the Baofeng radio that came to our rescue—its signal cutting through the blizzard like a beacon, connecting us to the outside world. It's not only me; countless others have had a similar experience with this humble device. Time and again, Baofeng radios have proven their worth in the most challenging situations.

In this chapter, we will delve into real-life stories of how Baofeng radios have proven their mettle in emergency situations, providing a lifeline when all other means of communication failed. We will explore how these radios have added a layer of safety and convenience to outdoor activities like hiking, camping, and geocaching (a kind of treasure hunt activity using a GPS).

From coordinating community marathons to orchestrating airshows, Baofeng radios have played a pivotal role. We will also touch upon the world of amateur radio operations, discussing nets, relays, and the etiquette in practice. You will discover the versatility of Baofeng radios through various case studies and scenarios.

BAOFENG RADIOS USED IN EMERGENCY SITUATIONS

We are all highly dependent on technology for communication. We tend to realize our dependency when natural disasters and other unforeseen circumstances disrupt our usual channels of communication. In such circumstances, Baofeng radios have proven themselves to be a lifeline. There are plenty of thrilling stories where a simple Baofeng radio came to save the day.

Alden Summer, an avid hiker, has one of these interesting stories. He was on a challenging journey along the Long Trail in Vermont. This trail, spanning 273 miles, traverses the state's highest mountains, offering breathtaking views but also presenting formidable challenges.

One day, while navigating this rugged terrain, Alden's blood sugar levels plummeted dangerously low. He lost consciousness and began to convulse in a seizure. The situation was dire, and immediate medical attention was needed. But the remote location of the trail meant that there was no cell reception, cutting off the primary means of communication with the outside world.

However, Alden was no amateur, and he had prepared himself for situations like this. He had brought along his Baofeng radio, a compact yet powerful device capable of connecting to local repeaters. Despite his deteriorating condition, Alden managed to tune into a local repeater on Mt. Greylock and broadcast a distress signal.

This call was picked up by two men, Ron Wonderlick and Matthew Sacco. Matthew, recognizing the urgency of the situation, grabbed his radio gear and drove to the Long Trail's parking lot, where he met the Incident Command Leader for the Search and Rescue (SAR) operation.

Matthew first tried to connect to the Mt. Greylock repeater using his handheld radio. He then tried using the mobile radio in his vehicle, which also failed. So, Matthew decided to get creative. He quickly assembled a J-pole antenna and attached it to his handheld radio. Using a fishing pole and a heavy sinker, he cast a line over a tree branch 20 feet in the air and hoisted his newly constructed antenna. This improvisation worked, and Matthew was able to establish a connection with the Mt. Greylock repeater, providing the SAR operation with much-needed outside communication.

The next challenge was to locate Alden. Given the trail's terrain, the only viable option was to send a helicopter. However, the SAR team had trouble communicating with the helicopter due to the poor range of their stock antennas. Once again, Alden's Baofeng radio came to the rescue. Alden had upgraded his antenna with an aftermarket version, which significantly improved its range. He lent his radio to the rescuers, enabling them to establish a reliable communication link with the helicopter, (Tate, 2021).

Thanks to Alden's Baofeng radio and Matthew's quick thinking, the helicopter was able to locate Alden, airlift him to safety, and undoubtedly save his life.

But Alden's experience is not the only one. Maya and Sarah, two friends, share a similar story where their Baofeng radio came out on top in the midst of grave circumstances.

Maya and Sarah were carving their way down a pristine black diamond run in the heart of the Canadian Rockies. The wind whipped around them, exhilaration coursing through their veins as their excitement knew no bounds. However, their joy was short-lived. A sudden, blinding snowstorm came down, engulfing the mountain in a blackout. Lost and disoriented, they stumbled off the marked runs. Their phones became useless in the blizzard, and panic started to grow.

But Maya had a device up her sleeve, The Baofeng UV–5R. Despite the harsh conditions, she managed to tune into a local repeater. There was hope. The receiver, John, with years of experience navigating the treacherous slopes, listened intently to Maya's panicked plea. Using his own Baofeng, he instructed them to stay put, conserving their energy. He then relayed their location to the ski patrol, pinpointing them on the map with surprising accuracy, thanks to the detailed description Maya provided through the radio.

Each minute felt like a lifetime. Soon the sound of spinning rotors came from far away. A rescue helicopter emerged from the swirling snow. One by one, Maya and Sarah were hoisted into the warm cabin of the helicopter. They came back to the ski lodge, safe and sound. A small Baofeng radio gave these young girls hope and became their guardian angel.

These are just a couple of examples of how Baofeng radios, despite their humble appearance, are in fact unsung heroes, proving their value in countless situations where reliable communication is essential. If you are a hiker, camper, or an explorer, it is imperative that you keep a Baofeng radio with you. You never know when it might just become your own life saver.

LEVERAGING YOUR BAOFENG FOR OUTDOOR ACTIVITIES

Baofeng radios have become a popular choice among outdoor enthusiasts. Whether you're hiking, camping, or geocaching, a Baofeng radio can enhance your experience by providing reliable communication and emergency support. In areas where cell service is unreliable or non-existent, Baofeng radios provide a dependable means of communication.

I'm sure pretty much every one of you reading this has been camping, hiking or geocaching. They are all fun activities that bring you close to nature and offer you a break from the hustle and bustle of everyday life. However, if you are not well equipped, they also present some unique challenges. This is where Baofeng radios come into play.

Baofeng radios operate on VHF (Very High Frequency) and UHF (Ultra High Frequency) bands, which allow for clear signal transmission over long distances. This makes them an ideal tool for coordinating with your group, whether you're setting up camp, planning hikes, or simply checking in on each other's well-being. If all members of your group have a Baofeng radio with them, you can rest assured that they are *not* going to get irretrievably lost—if they do, they will be able to trace themselves back to you.

The thrill of being out in nature comes at a price. Safety is never guaranteed and unexpected situations can arise, from sudden weather changes to medical emergencies. But just like Alden, if you have a Baofeng radio with you, you can access emergency frequencies, providing a critical lifeline in such situations. They can also receive NOAA weather alerts, keeping you informed of potential hazards and allowing you to take necessary precautions. This feature can be particularly useful when you are touring in areas prone to severe weather conditions.

Here is how you can call an emergency frequency using a Baofeng radio:

1. **Turn on the radio:** Make sure your Baofeng radio is charged and turned on.
2. **Enter Frequency Mode:** Press the VFO/MR button to enter Frequency Mode. In this mode, you can manually input the frequency you want to use.
3. **Input the Emergency Frequency:** The global distress frequency is 156.8 MHz. Use the keypad to input this frequency.
4. **Transmit Your Distress Call:** Press the PTT (Push-To-Talk) button, usually located on the side of the radio, to transmit your distress call. Speak clearly and calmly, stating your situation and location.
5. **Release the PTT Button to Receive:** After you've transmitted your message, release the PTT button. This will allow the radio to receive transmissions, so you can hear any incoming responses, (Nightingale, 2023).

Remember, it's important to only use emergency frequencies in genuine emergencies. Misuse can interfere with critical communications—and may be illegal.

Beyond the crucial practical benefits, Baofeng radios can also enhance your overall outdoor experience. For instance, they can be used for entertainment, such as listening to FM radio or communicating with other radio users. Some people even use them for radio orienteering, a fun and challenging activity that involves navigating with a map and radio receiver.

COORDINATING EVENTS WITH BAOFENG RADIOS

Baofeng radios, those unassuming little workhorses of the communication world, might not seem like the most glamorous tools for event coordination. But don't underestimate their power! From bustling community marathons to awe-inspiring airshows, these affordable and reliable radios can be the difference between a smooth operation, and chaos!

But I know what you are thinking'—*don't we have phones to communicate with each other? Why do we need Ham Radios during such events?*

Cellular networks can get overloaded in large crowds, especially outdoors. Baofeng radios operate on dedicated frequencies, offering instant, uninterrupted communication between organizers, volunteers, and security personnel. Think of it as a private phone line for your event team, free from dropped calls and frustrating network delays.

Unlike walkie-talkies, Baofeng radios often have duplex communication, allowing for two-way conversations without switching channels. This makes it easy to relay updates, issue instructions, and troubleshoot problems quickly and efficiently, like a race organizer instantly communicating with a marshal at a critical turn, ensuring the safety and flow of the event.

With dedicated channels for different teams, organizers can focus on specific tasks without getting bogged down in irrelevant chatter. Imagine security having a dedicated channel to report suspicious activity, while medics have another for coordinating emergency responses. This clear channel segmentation keeps everyone on their toes and focused on their specific roles.

Examples in Action:

- **Community Marathon:** Imagine a marathon manager using a Baofeng radio to communicate with marshals at key checkpoints, providing real-time updates on runner progress and potential hazards. This ensures the safety and smooth flow of the race while allowing organizers to quickly respond to any issues.
- **Airshow:** Picture a team of air traffic controllers using Baofeng radios to coordinate the movement of planes in the airshow, ensuring a safe and spectacular display. The clear communication and instant response time provided by these radios can be crucial in preventing accidents and ensuring a successful event.

Baofeng radios can be used for countless other events, from concerts and festivals to sporting events and conferences. Their ability to provide instant, clear, and efficient communication makes them invaluable tools for any organizer who wants their event to run smoothly, and safely.

So, the next time you're planning an event, don't underestimate the power of the little Baofeng radio. It might just be the secret weapon you need to turn your event from *good* to *great*.

AMATEUR RADIO OPERATIONS

Nets

An amateur radio net, or simply Ham Net, is an "on-the-air" gathering of amateur radio operators. Most nets convene on a regular schedule and specific frequency, and are organized for a particular purpose, such as relaying messages, discussing a common topic of interest, in severe weather (for example, during a Skywarn activation), emergencies, or simply as a regular gathering of friends for conversation.

Nets operate more or less formally depending on their purpose and organization. Groups of nets may organize and operate in collaboration for a common purpose, such as to pass along emergency messages in time of disaster. One such system of nets is the National Traffic System (NTS), organized and operated by members of the American Radio Relay League (ARRL) to handle routine and emergency messages on a nationwide and local basis, (Ham Radio Prep, n.d.).

The purpose of Baofeng radios in amateur radio nets is to facilitate communication among amateur radio operators. Their frequency coverage ensures you can find and join local nets. Programming features let you customize settings for specific nets, while their audio quality allows for clear communication.

Joining amateur radio nets can be an enriching experience, offering opportunities to connect with fellow hams, share information, and participate in various activities. Finding and joining them, however, might seem daunting at first. Here are some helpful tips to get you started:

- **ARRL Net Directory:** This comprehensive database lists hundreds of nets, categorized by frequency, region, and topic. You can search by location, band, or even net name to find ones that pique your interest.
- **Local Amateur Radio Clubs:** Most areas have active amateur radio clubs that organize regular nets. Check their websites, forums, or social media pages for net schedules and information.
- **Online Forums and Communities:** Many online forums and communities dedicated to amateur radio discuss local nets and offer tips for joining. Engage with members and ask for recommendations in your area.

Joining nets is a great way to learn, connect, and contribute to the vibrant amateur radio community. Begin with nets operating in your area to minimize signal issues and familiarize yourself with local hams. Then you can explore nets focused on specific topics, like emergency communications, technical discussions, or even casual social gatherings.

Relays

Think of a relay station as a high-powered radio repeater perched atop a hill or building. It receives signals from hams within its local area, amplifies them significantly, and then retransmits them to a wider range. This effectively extends the range of communication, allowing hams to talk to each other across distances they 'couldn't normally reach.

The reliability and ease of use of Baofeng radios make them popular choices for relay stations. Baofeng models can be easily programmed with repeater offsets, CTCSS tones, and auto-repeater functionality, simplifying relay operation. Their rugged construction ensures smooth relay operation, even in challenging conditions. Baofeng's interface is also very smooth, making it easy to use for monitoring and control.

Setting up a Baofeng-powered relay requires careful planning and consideration. Here are some key points:

- **Licensing and Regulations:** Understand and comply with local regulations regarding amateur radio repeater operation. Obtain necessary licenses and permits before setting up your relay.
- **Frequency Selection:** Choose frequencies designated for repeater use and avoid interfering with existing communication channels.
- **Equipment:** Consider factors like antenna height, power output, and cooling requirements to ensure adequate coverage and reliable operation.
- **Monitoring and Maintenance:** Regularly monitor your relay's signal strength and performance to identify and address any issues promptly.

Radio Etiquette

Just like any social setting, amateur radio has its own set of unwritten rules for respectful and efficient communication.

These principles ensure smooth, respectful, and efficient interactions, fostering a positive and productive environment for everyone. Here are some key points:

• Know Your Frequency:

- Always confirm the designated frequency for a net or repeater before transmitting. Listen carefully before talking to avoid interfering with others.
- Research and understand any specific operating procedures for your local nets or repeaters.

• Clear and Concise Communication:

- Keep your transmissions short and to the point. Avoid unnecessary chatter or rambling.
- Speak clearly and enunciate your words. Avoid background noise or microphone issues.
- Use your call sign consistently and correctly when identifying yourself or responding to others.

• Mind Your Power:

- Use the appropriate transmit power for the situation. Don't overpower weaker stations or exceed designated power limits.
- Adjust power based on the distance and terrain to ensure effective communication without unnecessary interference.

• Positive and Respectful:

- Be courteous and respectful in your interactions

with other hams. Avoid negativity, offensive language, or personal attacks.
• Pay attention to ongoing conversations before interrupting. Wait for appropriate breaks to transmit your own message.
• Respond promptly to calls and acknowledge other stations when appropriate.

Additional Tips:

- **Identify Yourself:** Walkie talkies don't always have caller ID and are meant to be picked up and used by anyone, so it's good etiquette to identify yourself when you start your conversation.
- **Learn The Lingo:** Certain words and phrases should be used when greeting, speaking to, and saying goodbye to another party. This is because some words used in everyday speech don't always transmit clearly over two-way radio waves.
- **Pause Before You Speak**: It's a good practice to pause for a second after pressing the PTT (press-to-talk) button on your two-way radio before you actually start to speak. This ensures your first word or two won't get cut off.

Baofeng radios offer a remarkably capable and affordable entry point into the world of amateur radio operations. With careful programming, proper etiquette, and a commitment to learning, Baofeng users can actively participate in nets, relays, and contribute to the vibrant community of amateur radio enthusiasts. And, of course, it's not just about the equipment, but about responsible, respectful and meaningful communication that makes amateur radio a truly rewarding and enriching experience.

9

NAVIGATING THE LEGAL LANDSCAPE

"Radio is the theater of the mind."

~ *Orson Welles*

A new piece of tech comes with new excitement, but it also brings you a new responsibility. With a Baofeng radio, you have the power to reach out to people far and wide, but you also have the responsibility to do so in a manner that respects the laws and ethics of radio communications. Often we get carried away, and we forget how there are actually some legal implications that come with this small piece of tech.

While experimenting with different frequencies, a friend of mine once stumbled upon a conversation between two operators. Overcome by curiosity and excitement, he decided to jump in and introduce himself. To his surprise, the conversation abruptly ended, and a few days later, he received a notice from the local communications authority.

He was in violation of several regulations—he had intruded on a licensed frequency and made an unauthorized transmission. He was facing hefty fines and the potential confiscation of his equipment. His excitement quickly turned into regret.

That incident was a wake-up call for him. He realized that operating a Baofeng radio wasn't just about the *thrill* of communication; it was also about respecting the laws and regulations that govern its use.

As a licensed user of Baofeng radios, I've navigated the complex landscape of radio communication laws, regulations, and ethics myself. Initially it might feel overwhelming. But don't worry, in this chapter, we are going to learn about these boundaries and how to respect them. From understanding the laws and regulations to obtaining a license for amplified use, we will cover it all. By the end of this chapter, you will not only understand the radio spectrum and Baofeng's place within it, but also appreciate the ethical considerations and best practices for responsible use.

LEGAL LANDSCAPE OF RADIO COMMUNICATION

Radio communication operates within a structured and regulated environment. These laws and regulations are designed to ensure orderly use of the radio spectrum, prevent harmful interference, and promote efficient use of resources.

The radio spectrum is a part of the electromagnetic spectrum, ranging from 3 kHz to 300 GHz, used for wireless communication. Different parts of the spectrum are allocated for different uses, such as AM/FM radio broadcasting, television broadcasting, mobile networks, satellite communication, and amateur radio, to name the main uses.

In most countries, the use of the radio spectrum is regulated by a government agency. In the United States, it's the Federal Communications Commission (FCC), in the United Kingdom, it's Ofcom, and in India, it's the Department of Telecommunications. These agencies are responsible for issuing licenses, setting technical standards, and enforcing regulations.

Baofeng radios, popular among amateur radio enthusiasts, are capable of transmitting on a wide range of frequencies. However, not all of these frequencies are legal to use without a license, and some are not legal to use at all due to allocation to other services. The process of acquiring a license involves studying for and passing an examination that tests your understanding of radio theory, operating practices, and legal regulations. Once you pass the exam, you're issued a call sign that you must use to identify your transmissions. But even *after* acquiring your license, you need to understand which frequencies you're allowed to operate on.

Failure to comply with any radio communication laws and regulations can result in penalties, including fines, confiscation of equipment, and in some cases, imprisonment. Therefore, it's clearly better keep yourself abreast of the law and learn the rules. Beyond the legal aspect, there are also ethical considerations in radio communication. These include respecting the rights of other users, avoiding unnecessary interference, and using the spectrum in a way that benefits the community. So, it's not just about following the rules, but also about being a responsible and ethical member of the radio community.

UNDERSTANDING HOW TO OBTAIN A LICENSE FOR AMPLIFIED USE

When it comes to amplifying the radio's power beyond its standard output, navigating the licensing landscape can be daunting. Before we talk about the licensing process, t's crucial to understand the governing bodies and regulations impacting radio use in your region. In most countries, two primary authorities regulate radio communications:

- **The Federal Communications Commission (FCC):** Responsible for regulating radio spectrum usage in the United States.
- **International Telecommunication Union (ITU):** Sets international standards and regulations for radio communication, which most countries adhere to, (Guide and Romanchik, 2023).

The type of license you need depends on your intended use of the amplified Baofeng radio. Here are the two primary options:

- **Amateur Radio License (Ham License):** This internationally recognized license grants access to a wider range of frequencies and higher power output compared to license-free options. Obtaining a Ham license requires passing a written examination covering radio theory, regulations, and operating procedures. There are three classes of Ham Radio licenses in the US: Technician; General; and Extra. Each class grants access to different frequencies and privileges. The Technician license is the easiest to get and only requires passing a 35-question exam on basic radio theory and regulations.

- **General Mobile Radio Service (GMRS) License:** This license allows operation on specific frequencies designated for private, two-way communications. In the US, GMRS licenses are issued by the FCC without requiring an examination. If you only plan to use your Baofeng on GMRS frequencies (typically around 151.625 MHz to 154.675 MHz) and with the radio's standard power output (around 5 watts), you can use it *without* a license in most countries.

While the GMRS license does not require an exam, here are some resources that can help you get started on preparation if you are going to sit for the exam to get your Ham Radio license.

- **Official FCC Study Guides**: The FCC provides free downloadable study guides covering radio theory, regulations, and operating practices.
- **Amateur Radio Organizations**: Numerous amateur radio organizations offer online resources, study courses, and practice exams. Popular options include the American Radio Relay League (ARRL) and the National Association for Amateur Radio (NAAR).
- **Licensed Ham Mentors**: Connecting with a licensed Ham can be invaluable. They can answer

your questions, provide practical guidance, and even conduct mock exams.

Once you've conquered the exam, it's time to file the necessary paperwork. The specific forms and procedures vary depending on your location. In the US, you'll need to complete and submit the FCC Form 605 application to the FCC. After completing these forms, you need to send the payment. The cost for the license is $70, but it was reduced to $35 for 10 years as of April 19, 2022, (Chen, 2021).

By diligently studying, filing the necessary paperwork, and operating responsibly, you can unlock the full potential of your amplified Baofeng radio and become a valuable member of the global radio community! I know this process might feel like a bit of a hassle, but the journey is as rewarding as the destination. So, enjoy it, stay informed and enjoy a decade of seamless communication with the flexibility of your license.

UNDERSTANDING THE RADIO SPECTRUM AND BAOFENG'S PLACE WITHIN IT

Radio waves have different frequencies, and by tuning a radio receiver to a specific frequency, you can pick up a specific signal. Frequencies are often grouped in ranges called bands. For example, all frequency modulated (FM) radio stations transmit in a band of frequencies between 88 MHz and 108 MHz. This band of the radio spectrum is used for no other purpose other than FM radio broadcasts. Similarly, amplitude modulated (AM) radio is confined to a band from 535 kHz to 1,700 kHz. (Brain 2023)

For the purposes of understanding Baofeng radios, we'll focus on the high-frequency and ultra-high-frequency bands (HF and UHF, respectively). These bands, roughly spanning 3 MHz to 3 GHz, host a bustling marketplace of users:

- **Amateur Radio (Ham):** Licensed enthusiasts communicate on designated frequencies, fostering communication, technical exploration, and emergency response networks.
- **Marine VHF:** Boats and coastal authorities rely on dedicated channels for safety and navigation.
- **Public Safety:** Police, fire, and emergency medical services operate on secure channels to coordinate their vital work.
- **Business and Industrial:** Companies utilize specific frequencies for private communication within their operations.
- **Military and Government:** Secure channels for confidential communication and mission-critical operations.

Within this crowded landscape, Baofeng radios offer a multi-band, affordable option. Many models cover both the VHF and UHF amateur radio bands (144–148 MHz and 420–450 MHz), allowing users to engage in licensed amateur radio activities. However, their versatility extends beyond this specific realm. Some models cover additional frequencies used for Marine VHF, GMRS (General Mobile Radio Service), and even business/industrial applications.

ETHICAL CONSIDERATIONS AND BEST PRACTICES FOR RESPONSIBLE USE

Beyond the technical aspects and legal boundaries lies a crucial layer of ethical considerations that need to guide the actions of every Baofeng user. Ignoring these considerations can not only land you in legal trouble, but also disrupt vital communication channels and put others at risk.

The radio band is like a busy highway, with each frequency being its own lane. Users of Baofeng use lanes that are set aside for amateur radio, naval VHF, GMRS, and other services. If you're not allowed to be in that lane, transmitting is like turning into oncoming traffic—which could cause chaos and put other users in danger. So, it is very important we know which frequencies are allowed in our license and make sure to *always* stick to them. Consider it your ethical duty to respect the established order of the spectrum and avoid encroaching on others' channels.

Baofengs, when misused, can become instruments of unintentional interference, jeopardizing critical communication channels and potentially putting lives at risk. *Always* check for existing activity on a frequency *before* transmitting, and prioritize emergency communications over casual conversations. What happens if your crackling voice interrupts a doctor coordinating emergency medical aid, or a garbled message disrupting a maritime distress call? You could be putting people's lives in danger. While these devices can be great sources of entertainment, the airwaves are not your playground; responsible use means ensuring your transmissions don't become someone else's nightmare.

Again, the ability to listen to conversations on open frequencies comes with the power to invade privacy. Eavesdropping on private conversations, whether intentionally or unintentionally, is not only *unethical* but also potentially *illegal*. Treat the airwaves with respect and avoid listening to conversations not intended for you.

Additionally, maintain proper etiquette while transmitting. Avoid using offensive language, refrain from dominating channels with lengthy monologues, and be mindful of the volume of your transmissions. The airwaves are a shared space, and courtesy goes a long way in fostering a positive and respectful community.

While following regulations is essential, responsible Baofeng use extends beyond mere compliance. It's about understanding the spirit of the law, the ethical considerations that underpin it, and the potential impact of your actions on the broader community of users. Be a responsible Baofeng user, someone who not only adheres to the rules, but also proactively promotes responsible and ethical practices within the airwaves.

With the knowledge gained from this chapter, you've taken a critical step towards responsible Baofeng use. You've dipped your toes into the stream of legalities in the radio spectrum, learned the language of permits and licenses, and gained a deeper appreciation for the ethical considerations that guide every transmission. Yet, your journey is far from over.

In the next chapter, "Elevating Your Baofeng Experience," we'll look into ways of boosting your overall Ham Radio experience with innovative new habits and practices.

10

BONUS CHAPTER
ELEVATING YOUR BAOFENG EXPERIENCE

"The air is always full of voices. All we have to do is learn how to listen."

~ *Jeanette Winterson*

Having a Baofeng radio doesn't only make you a radio enthusiast, it makes you an explorer—an explorer of airwaves. While I always find radio communication exciting, participating in amateur radio competitions and getting myself involved with the wider radio community led me to fall in love with the device. The concept of engaging with a community from miles away, without any visual contact, is thrilling. It just makes the experience of using radios more surreal!

This chapter promises to be an exciting introduction to the possibilities that lie beyond the basics of Baofeng use. By engaging with the community, participating in competitions, and embracing advanced technologies, you'll transform your radio from being merely a tool into a springboard for exploration and learning.

My journey with Baofeng radio has been a pathway of continuous learning. From clubs, forums, and social media to online resources, courses, and local workshops, every step has been a lesson, and every interaction a new discovery. In this chapter, you will learn how you can pursue this approach of continuous learning.

ENGAGING WITH THE WIDER RADIO COMMUNITY

"Gone are the days of solitary tinkering; radio enthusiasts from all walks of life gather in clubs, forums, and social media to share knowledge, swap stories, and celebrate the shared passion that binds them together. Engaging with this community is not just about tips and tricks; it's about *finding your tribe*, a group of kindred spirits who will fuel your radio journey with inspiration and support.

Radio clubs are a great place to start. They provide a platform for enthusiasts to meet, exchange ideas, and organize events. Here's how you can get involved:

- **Local Amateur Radio Clubs**: Look for a local club in your area. These clubs often hold regular meetings and events, which are excellent opportunities to learn and network.
- **Special Interest Clubs:** These clubs focus on specific aspects of amateur radio, such as emergency communication, contesting, or digital modes. Joining these clubs can deepen your understanding of these areas.

Online forums are another excellent resource. The internet pulsates with the chatter of online forums and virtual hubs, where radio enthusiasts from every corner of the globe converge. Imagine a bustling marketplace of ideas, where technical debates mix with friendly banter, and troubleshooting tips fly faster than electrons. They offer a platform to ask questions, share experiences, and learn from others. Here are some popular ones:

- **QRZ Forums:** QRZ.com is one of the most visited websites in the amateur radio world. Their forums cover a wide range of topics.
- **Reddit's/r/amateurradio**: This subreddit is a vibrant community of amateur radio enthusiasts. It's a great place to ask questions and share your experiences.

It is worth repeating: it is important you always respect the rules of the forum and maintain a positive and respectful attitude. You can also consider leveraging Facebook groups, Twitter hashtags, and YouTube channels. Social media provides an instant connection to the vibrant pulse of the radio community. But it's obviously a two-way street. You need to engage with others, share your knowledge, and be a positive force in the online community as well.

There will be a place for you in the radio community, whether you want to meet with people in a local club, on an online forum, or in a group chat through social media. Embrace the welcoming spirit, ask questions, learn from others, and share your own knowledge. You'll soon discover that your Baofeng isn't "just a radio"; it's a key that opens doors to a lively world of friendship, learning, and endless opportunities on the airwaves.

AMATEUR RADIO COMPETITIONS

So, you've mastered the basics, conquered the legalities, and are itching to push your Baofeng beyond casual QSOs. It's time to embrace the competitive spirit! Amateur radio competitions offer a thrilling arena to test your skills, hone your techniques, and experience the exhilaration of friendly rivalry on the airwaves.

Just like there are endless ways to use your Baofeng, there's a diverse landscape of contests catering to every skill level and interest. Some popular competitions include:

- **Field Day:** This is an annual amateur radio contest that encourages emergency communications preparedness among amateur radio operators. Participants set up temporary transmitting stations in public places to demonstrate 'Ham Radio's science, skill, and service to communities. The goal is to make as many contacts as possible within a 24-hour period. It's like camping with a side of radio magic!
- **Sprint Contests:** These are intense bursts of activity for the fast-paced ham. They are short contests, sometimes as brief as 30 minutes, that demand lightning-fast contact-making and efficient log-keeping. The objective is for North American stations to contact as many radio amateurs as possible.
- **DX Contests:** These contests aim for contacts with stations in far-flung corners of the globe. The CQ World Wide DX Contest, for example, is the largest Amateur Radio competition in the world. Over 35,000 participants take to the airwaves on

the last weekend of October (SSB) and November (CW) with the goal of making as many contacts with as many different DXCC entities and CQ Zones as possible.

- **Theme-based Contests:** These contests focus on a specific skill or technique. Examples include CW (Morse code) challenges and QRP (low-power) epics. Participants dive into a specific skill or technique, pushing out their comfort zone and discovering hidden capabilities of their equipment, (Cloud Radio, 2019).

No two competitions are alike. Whether you choose a pulse-pounding sprint lasting a mere 30 minutes or a grueling 24-hour Field Day extravaganza, each contest offers a unique flavor in the range of challenges. There are contests for DX enthusiasts seeking far-flung contacts, CW aficionados testing their Morse code prowess, and QRP masters pushing the boundaries of low-power communication. No matter your skill level or interest, there's a competition out there waiting for you to unleash your inner radio enthusiast.

Imagine—you are scanning the bands and hearing a symphony of call-signs moving around like fireflies at night. You feel a rush of excitement as a voice cuts through the noise. It's a new contact. Every QSO you make adds to your score, and every exchange you complete is a hard-fought win. Your brain is busy making plans, your blood is pumping, and your Baofeng is humming with the energy of your determined drive. It's like playing chess in your head at fast speed! This is exactly what you can expect from these amateur radio competitions—thrill, joy and excitement.

But there is a strong sense of community burning brightly beneath the competition. These aren't just competitions that you fight by yourself; they're tours that you take with other radio enthusiasts. You'll share tips, celebrate wins, and offer support if you're feeling down. Even your competitors are part of the vibrant radio community, and share a similar passion for the airwaves.

Trophies and awards are certainly sweet, but the true reward lies in the journey itself. Competitions push you to explore new facets of your radio, experiment with techniques, and expand your knowledge of the radio spectrum. You'll develop invaluable skills, like time management, strategic thinking, and quick decision-making —skills that translate far beyond the airwaves and enrich your life in countless ways.

CRAFTING A PATH FOR CONTINUOUS LEARNING

The teachings of this book have certainly sparked a fire of interest within you. But this passion needs fuel, and in the case of your Baofeng radio, the fuel is continuous learning. The internet is a treasure trove for the curious radio explorer. Here are some sources to choose from:

- **Baofeng Official Website**: This is obviously the first place to visit for any Baofeng radio user. The website provides comprehensive information about various Baofeng radio models, accessories, and new arrivals. It also features a blog section where you can find articles on topics like applying for a GMRS license, programming Baofeng UV–5R series with programming software, and more.

- **Online Learning Platforms**: Websites like Udemy or Coursera offer a wealth of radio-related courses, from introductory tutorials to advanced techniques in digital modes and satellite communication. Many courses are even free, a testament to the generous spirit of the radio community.
- **E-Books and Manuals:** Download comprehensive guides on Baofeng models, antenna theory, and contest strategies. Learn at your own pace, revisit key concepts, and build your personal library of radio knowledge.
- **Forums and Communities**: Join online forums dedicated to Baofengs and contest enthusiasts. Ask questions, share experiences, and get valuable feedback from seasoned hams. You'll find troubleshooting tips, contest debriefs, and a constant flow of new ideas to keep your learning flame burning bright.
- **Blogs and YouTube Channels**: Follow expert radio bloggers and YouTubers who share their knowledge and adventures on the airwaves. Watch tutorials, listen to contest highlights, and get inspired by the stories of radio enthusiasts from around the world.

It is important you remember that gaining knowledge is only *half* the battle. To truly hone your skills and conquer future applications, you need to *practice,* and there is no substitute for real-world experience. Many online platforms and organizations host regular practice contests. These are perfect for testing your radio knowledge, trying out new equipment, and building experience.

Don't be afraid to reach out to experienced hams, whether online or in your local club. Seek guidance, ask questions, and participate in mentoring programs. Their insights and feedback can be invaluable in your quest for learning everything there is to learn about Ham Radios.

The path of continuous learning through competitions is not a sprint, but a marathon. So, as you navigate online resources, attend workshops, and practice your skills, remember to celebrate the joy of learning, the thrill of discovery, and the satisfaction of improving your skills. Pay it forward by helping others learn. Answer questions on forums, mentor newcomers in your club, and share your experiences with the wider radio community. The act of teaching not only solidifies your own understanding but also fosters a sense of collaboration.

As I hope I have shown you in this book, the world of radio is vast and ever-evolving. Embrace new technologies, experiment with different modes and techniques, and always keep your mind open to new possibilities.

CONCLUSION

"Radio affects most intimately, person-to-person, offering a world of unspoken communication between writer-speaker and the listener."

~ *Marshall McLuhan*

I hope that by now, the sounds of your Baofeng no longer appear like whispers in the dark. They're an orchestra of possibilities waiting for the symphony to be conducted. You've learned the instrument, mastered its controls, and now it's time to venture on your own radio journey.

Your Baofeng is not "just a radio"; it's a tool for forging connections, igniting adventures, and pushing the limits of possibility. It can bridge the gap between loved ones who find themselves miles apart, it can be your lifeline in an emergency, and guide you through unknown terrain. It grants you a voice within the radio community—a fellowship bound by shared passion.

Reading this book is only just the beginning. Take its lessons, its insights, and its stories as springboards for your own journey. The whole idea of understanding the Baofeng radio is to apply it in real life circumstances. So, pick up your Baofeng, tune in to the sea of possibilities, and compose your *own* radio story. Share your knowledge, hone your skills, and become a part of the community that breathes life into these electronic wonders.

Take a deep breath, adjust your antenna, and turn up the volume. Your 'Baofeng symphony is ready to begin. And you, the conductor, are about to guide your own vision of a masterpiece.

REFERENCES

Access Agility. 2018. *WiFi Antenna Types*. AccessAgility. https://www.ac-cessagility.com/blog/wifi-antenna-types.

Amateur Radio Stack Exchange. 2015. uv 5r—*Storage of a Baofeng UV–5R*. Amateur Radio Stack Exchange.

https://ham.stackexchange.com/questions/3488/storage-of-a-baofeng-UV-5R.

Baofeng. n.d. *About Us*. Baofeng. Accessed November 27, 2023. https://www.baofengradio.com/pages/about-us.

Baofeng. n.d. *Baofeng UV-5R Manual*. Baofeng Tech. Accessed November 28, 2023. https://baofengtech.com/wp-content/uploads/2020/10/BaoFeng_UV-5R_Manual.pdf.

Baofeng Manual. 2023. *Support and Help—BaoFeng*. Baofeng Tech. https://baofengtech.com/support/.

Baofeng Manual. n.d. *BAOFENG UV-5R 5Watt UHF/VHF Radio*. Baofeng.

Accessed December 17, 2023. https://www.baofengradio.com/prod-ucts/catalog-UV-5R.

Baofeng Manual. n.d. *BAOFENG UV-9G GMRS IP67 Waterproof Radio*. Baofeng.

Accessed December 15, 2023. https://www.baofengradio.com/products/uv-9g.

Baofeng Manual. n.d. *BF–F8HP User Manual*. BaoFeng Tech. Accessed December 15, 2023. https://baofengtech.com/wp-content/uploads/2020/09/BF-F8HP_Manual.pdf.

Baofeng Tech. n.d. *Nagoya NA–701C - BaoFeng Radios*. Baofeng Tech. Accessed December 6, 2023. https://baofengtech.com/product/nagoya-701c/.

Baofeng UV–5R Manual. n.d. *Baofeng Tech*. Accessed December 6, 2023. https://baofengtech.com/wp-content/uploads/2020/10/BaoFeng_UV–5R_Manual.pdf.

The Beaten Path. n.d. *Review of Abbree AR-152 Tactical GMRS Antenna For Long Range*. Woof The Beaten Path. Accessed December 6, 2023.

https://woofthebeatenpath.com/can-tactical-gmrs-antennas-increase-handheld-range/.

Bimmer. 2021. *Properly set DCS or CTCSS on UV–5R: r/Baofeng*. Reddit. https://www.reddit.com/r/Baofeng/comments/mrxony/properly_set_d-cs_or_ctcss_on_uv5r/.

Brain, Marshall. 2023. *How the Radio Spectrum Works* | HowStuffWorks. Electronics | https://electronics.howstuffworks.com/radio-spectrum.htm.

Brittanica. n.d. *Radio wave | Examples, Uses, Facts, & Range.* Britannica. Accessed December 17, 2023. https://www.britannica.com/science/radio-wave.

Chen, Jackson. 2021. *How to Apply for a GMRS License?* Baofeng. https://www.baofengradio.com/blogs/news/gmrs-license.

Cloud Radio. 2019. Radio Contest Ideas: 8 Hot-Practical Ideas – Services. CloudRadio. https://www.cloudrad.io/radio-contest-ideas.

Commscope. n.d. *Omnidirectional Antennas.* CommScope. Accessed December 17, 2023. https://www.commscope.com/product-type/antennas/base-station-antennas-equipment/base-station-antennas/omnidirectional/.

Crider, Thomas. 2019. *Building a better antenna for my Baofeng: r/amateurradio.* Reddit. https://www.reddit.com/r/amateurradio/comments/b7uw93/building_a_better_antenna_for_my_baofeng/.

Fisher, Tim. 2023. *What Is Firmware? Lifewire.*
https://www.lifewire.com/what-is-firmware-2625881.

Guide, Step, and Dan Romanchik. 2023. *Step-by-Step Guide to Getting Your 10-Year GMRS License: Easy Application Process Explained - BaoFeng Radios.* Baofeng Tech. https://baofengtech.com/step-by-step-getting-a-gmrs-license/.

Ham Radio Prep. n.d. *The Ultimate Baofeng Guide for Ham Radio.* Ham Radio Prep. Accessed December 16, 2023. https://hamradioprep.com/baofeng-vhf-and-uhf-handheld-radio/.

LiGo. 2021. *What is VOX on a Two-Way Radio?* LiGo Electronics.
https://ligo.co.uk/blog/vox-two-way-radio/.

Meacher, Nick. 2019. *New FCC Rules that Impact Preppers and New Hams.* Survival Dispatch. https://survivaldispatch.com/new-fcc-rules-that-impact-preppers-and-new-hams/.

Michael M. 2018. *How To Program Your Radio – Baofeng.* Baofeng. https://baofeng.zendesk.com/hc/en-us/articles/360002571551-How-To-Program-Your-Radio.

Morgan, Steve. 2023. \/. YouTube. https://thevideoanswers.com/how-do-you-use-a-baofeng-dual-band-fm-transceiver/.

Nightingale, Alex. 2023. *How to Use Baofeng Radios: A Quickstart Guide to the Keypad And Your Programming File — Aganz Store.* Aganz Store. https://www.aganzstore.com/blog/baofeng-radios-a-quickstart-guide-to-your-programming-file.

Nightingale, Alex. 2023. *How to Use Baofeng Radios: A Quickstart Guide to the Keypad And Your Programming File — Aganz Store.* Aganz Store.

https://www.aganzstore.com/blog/baofeng-radios-a-quickstart-guide-to-your-programming-file.

Olander, Travis. 2022. *The Ham Radio Guide: Navigating the UV-5R Menu.* AllOutdoor.com. https://www.alloutdoor.com/2022/12/21/guide-ham-radio-UV-5R-menu-programming/.

RadioReference. 2013. *How to program repeaters into a Baofeng UV–5R.* RadioReference.com Forums. https://forums.radioreference.com/threads/how-to-program-repeaters-into-a-baofeng-UV-5R.268176/.

Radtels. 2023. \/. YouTube. https://www.radtels.com/products/radtel-full-band-tactical-antenna-136-520mhz-for-amateur-rt-490-rt-470l-rt-470-rt-830-rt-890-ip68-rt-780-rt-659-rt-770-rt-760?variant=43591616987344.

Schweber, Bill. 2017. *Antenna Principles, Part 1.* DigiKey. https://www.digikey.com/en/articles/understanding-antenna-specifications-and-operation.

Sherman, Jon. n.d. *Programming Your Baofeng UV–5R Radio CHIRP Software Programming Guide.* iwillprepare.com. Accessed December 5, 2023.
http://www.iwillprepare.com/files/pdf/baofeng_%20uv5r_chirp_programming_guide.pdf.

Smith, Dan. n.d. *Beginners Guide.* CHIRP. Accessed December 5, 2023. https://chirp.danplanet.com/projects/chirp/wiki/Beginners_Guide.

Smith, Dan. n.d. *Download.* CHIRP. Accessed December 5, 2023. https://chirp.danplanet.com/projects/chirp/wiki/Download.

Stack Exchange. 2021. *Can Baofeng UV-5R HTQ scan and handle TDR (split) mode at the same time?* Amateur Radio Stack Exchange. https://ham.stackexchange.com/questions/20317/can-baofeng-UV-5R-htq-scan-and-handle-tdr-split-mode-at-the-same-time.

Stone, Alyssa *et al.* 2017. *When it Comes to Antennas, Size Matters.* Northeastern Global News. https://news.northeastern.edu/2017/08/23/when-it-comes-to-antennas-size-matters/.

Sturcke, James. 2010. *Haiti earthquake: survivors' stories | Haiti.* The Guardian. https://www.theguardian.com/world/2010/jan/14/haiti-earthquake-survivors.

Tate, Aden. 2021. *Real Life Survival Story: Baofeng, the Little Ham Radio That Could—Clatsop County, Oregon AuxComm.* Clatsop County AuxComm. https://clatsopauxcomm.org/277-real-life-survival-story-baofeng-the-little-ham-radio-that-could.

Wikipedia. n.d. *Baofeng UV–5R.* Wikipedia. Accessed November 27, 2023. https://en.wikipedia.org/wiki/Baofeng_UV-5R.

Woodford, Chris *et al.* 2022. *Radio and digital radio | How it works | AM and FM compared*. Explain that Stuff. https://www.explainthatstuff.com/radio.html.

Thanks for purchasing my book!

You can claim your free bonus by scanning the below QR-code.

Made in United States
Troutdale, OR
02/24/2024

17895110R00076